The Leopard Lily Project

Edzard Ernst

The Leopard Lily Project

A Story of Nazi Pseudoscience

Edzard Ernst
Exeter University
Exeter, UK

ISBN 978-3-032-24770-4   ISBN 978-3-032-24771-1   (eBook)
https://doi.org/10.1007/978-3-032-24771-1

This Springer imprint is published by the registered company Springer Nature Switzerland AG
The registered company address is: Gewerbestrasse 11, 6330 Cham, Switzerland

If disposing of this product, please recycle the paper.

To Danielle

Preface

The enemy must not only be defeated but destroyed

This book tells the bewildering tale of one of the most insidious, yet almost forgotten crimes against humanity planned by the Nazis during the Third Reich. The story begins in 1941 with the Austrian dermatologist, Adolf Pokorny. He wrote to Himmler stating that "the enemy must not only be defeated but destroyed." The letter suggests that an herbal remedy, Leopard Lily, should be used to secretly sterilize many millions of people. His method, Pokorny explains to Himmler, has the advantage that the regime can continue exploiting the victims as slave laborers and, at the same time, make sure these "Bolsheviks" cannot reproduce.

Himmler, who had long searched for such a solution, is enthusiastic about Pokorny's suggestion. At that stage, the Third Reich is facing the problem of the many millions who inhabit the occupied territories in the East. On the one hand, these people are badly needed as a workforce and must therefore be kept alive. On the other hand, Himmler wants them to disappear once the war is won and the territories can be fully "germanized."

Himmler sees himself as a gifted scientist and has a particular weakness for alternative medicine. He had already initiated pseudoscientific experiments on concentration camp inmates to investigate various brutal methods of mass sterilization. Yet, none of them proved to be suitable for his specific purpose. Pokorny's suggestion of employing an otherwise harmless herbal remedy therefore seems ideal to Himmler, and he decides to pursue this project as a matter of urgency. The Leopard Lily project thus becomes one of the most monstrous attempts at ethnic cleansing ever conceived.

To fully understand this complex story, we ought to know about some of the deeply immoral ideologies of the Nazis, their fascination with alternative medicine, particularly herbal medicine, Himmler's inclination toward pseudoscience, as well as the individuals involved.

The book connects the various strands of this complex plot and tells the disturbing tale of the Nazis' fanatical determination to commit genocide.

Exeter, UK   Edzard Ernst

Contents

1

Hitler and the Ideology of Race Hygiene

This chapter provides some context for understanding Hitlers ideology of racial hygiene. It was used to justify the systematic persecution, sterilization, and murder of large groups of individuals deemed to be "inferior." The German medical profession proved to be an essential instrument for implementing atrocities that are unprecedented in the history of medicine.

Countless volumes have been written about Hitler, his life, ideas, motivations, actions, etc. This chapter does not attempt to add to this wealth of information. Its aim is merely to provide some essential context for understanding the rest of this book.

The Genesis of an Ideology

After the failed "Beer Hall Putsch" (Box 1) in November 1923—an attempt by Hitler and the Nazi Party to overthrow the Bavarian government—Hitler was imprisoned in Landsberg, Bavaria.[1] During his incarceration from April 1, 1924, to December 20 of the same year, he studied the book by Fritz Lenz (Box 2), *Grundriss der menschlichen Erblichkeitslehre und Rassenhygiene* (Basic

---

[1] The Trial of Adolf Hitler: The Beer Hall Putsch and the Rise of Nazi Germany: King, David: Amazon.de: Books King D: The Trial of Adolf Hitler: The Beer Hall Putsch and the Rise of Nazi Germany. Pan (2018).

E. Ernst, *The Leopard Lily Project*, https://doi.org/10.1007/978-3-032-24771-1_1

Outline of Human Heredity and Racial Hygiene).[2] While imprisoned, Hitler dictated his own book, "Mein Kampf,"[3] to his fellow prisoner Rudolf Hess.[4]

Published almost a decade before the Nazis seized power, the book clearly outlined Hitler's concept of racial hygiene. This ideology would later provide the core justification for the barbaric policies of the "Third Reich," including involuntary sterilizations, the euphemistically named "euthanasia" programs, and the Holocaust.

In his book, Hitler advocated measures to prevent the "degeneration" of the German people, which he attributed to "inter-breeding" with groups he deemed to be "racially inferior," such as Jews, Slavs, and Roma. Hitler stressed the imperative to promote the reproduction of "racially valuable" individuals while preventing the propagation of those considered "inferior," presenting these measures as crucial for national survival:

> The result of all racial crossing is… always the following: lowering of the level of the higher race; physical and intellectual regression and hence the beginning of a slowly but surely progressing sickness. The prevention of the faculty and opportunity to procreate on the part of the physically degenerate and mentally sick, over the period of only six hundred years, would not only free humanity from an immeasurable misfortune, but would lead to a recovery which today seems scarcely conceivable. The State must set race in the center of all life. It must take care to keep it pure. It must declare the child to be the most precious treasure of the people. It must see to it that only the healthy beget children; that there is only one disgrace: to be sick and to bring children into the world; and one highest honor: to renounce doing so.

The concept of race hygiene was not Hitler's innovation; he merely adopted it and exploited it for his purpose. The idea of Aryan superiority had first been proposed by the French aristocrat and anthropologist, Joseph Arthur de Gobineau, in 1853.[5] In 1859 Darwin published his seminal book *On the Origin of Species*,[6] and the term "eugenics" was defined in 1883 by the Englishman, Francis Galton.[7]

---

[2] Grundriß der menschlichen Erblichkeitslehre und Rassenhygiene (2/2): by Fritz Lenz (English Edition) eBook: Lenz, Fritz: Amazon.de: Kindle Store Lenz F: Grundriß der menschlichen Erblichkeitslehre und Rassenhygiene. Lehmanns Verlag, (1921).

[3] Mein KampfWikipedia.

[4] The Truth About Rudolf Hess eBook: Douglas-Hamilton, James: Amazon.co.uk: Kindle Store Douglas-Hamilton J: The Truth About Rudolf Hess. Frontline Books (2020).

[5] An Essay on the Inequality of the Human Races—Wikipedia.

[6] On the Origin of Species—Wikipedia.

[7] Francis Galton—Wikipedia.

In the decades before the Nazis seized power, these concepts were popular in Germany and beyond. Alfred Ploetz had founded the German Society for Racial Hygiene in 1905[8] alongside with Wilhelm Schallmayer[9] who popularized the notion that the state should intervene to improve the "racial stock" through selective breeding and sterilization. Fritz Lenz (Box 2) became the editor of the *Archiv für Rassen- und Gesellschafts-Biologie* (Archive for Racial and Social Biology) in 1913, and in 1923 he was appointed to Germany's first chair in Racial Hygiene at the University of Munich. He explicitly called for the sterilization of the "most unfit third of the population" in 1931.

The Implementation of Nazi Policy

After Hitler was appointed Reich Chancellor on January 30, 1933, the Nazis speedily implemented their policy of "Gleichschaltung," the alignment of individuals and institutions with their ideology, and established a totalitarian dictatorship[10]:

- Within months, all political parties other than their own, the NSDAP, were abolished.
- Many basic rights were suspended.
- The police were transformed into an instrument of Nazi suppression.
- Opposition leaders were incarcerated.
- The rule of law was replaced by the "Führer principle," asserting Hitler's absolute authority.
- Extensive propaganda campaigns were crafted to indoctrinate the German public.

The Nazis perverted Social Darwinism into the notion of "race hygiene." It determined that Jews and other non-Aryans had to be defamed, disowned, and eventually eliminated. Initially, Jews were merely disenfranchised, marginalized, and often forced to emigrate. The "Law for the Restoration of the Professional Civil Service" excluded them from public office. The "Nuremberg Race Laws," enacted in 1935, stripped Jews of their German citizenship and prohibited them from marrying non-Jews. Concurrently, the systematic extermination of all European Jews was secretly planned, organized, and eventually put into action.

---

[8] Alfred Ploetz—Wikipedia.
[9] Wilhelm Schallmayer—Wikipedia.
[10] Nazi Germany—Wikipedia.

The sterilization of individuals considered "unworthy of life" had been a subject of discussion for several years, driven by eugenic concepts and the desire to alleviate perceived financial burdens on society. Voluntary sterilization had been legal since the early 1930s, and in 1933, the Nazis passed the law that provided the legal basis for the involuntary sterilization of disabled Germans (Chap. 2).

The outbreak of World War II, on September 1, 1939, with the invasion of Poland, provided convenient cover for the systematic murder of "unwanted individuals." Victims initially included disabled patients and others deemed "ballast existences." However, this category soon expanded to encompass anyone who did not conform to Nazi ideology. Overt utilitarianism defining certain individuals as less valuable than others now dominated all aspects of German life. The primary criterion for differentiating the "ballast" from the "valuable" was the ability to contribute to the advancement of the "Volk" (people) and to the war effort. Joseph Goebbels' propaganda proclaimed that Germany's destiny was to rule the world and asserted that the racially superior Aryan population would establish permanent dominion over Europe and beyond.

The Medicalization of Persecution

Racial hygiene had been converted into a medical issue. Throughout "Mein Kampf," Hitler referred to the Jewish race in medical terms, calling them "bacillus," "parasite," and "disease." Goebbels' propaganda relentlessly reinforced these medical analogies and propagated the idea that the "biological body of the German people" (*Volkskörper*) was diseased, and Hitler was the "healer" who would eradicate this "assault" on the nation's health once and for all. In this way, the "Jewish question" was reframed as a medical problem that demanded a medical solution.

The German medical profession was captivated by Nazi ideology.[11] A higher percentage of physicians, compared to any other profession, were members or sympathizers of the NSDAP. Before the "Third Reich," unemployment among doctors had been rampant. The Nazis promised to radically improve this situation and restore physicians' income, status, and honor. Most German doctors enthusiastically embraced these prospects and willingly followed Hitler and his ideas. Resistance among the German medical

---

[11] Lavik NJ. Tysk medisin under nazismen–samarbeid og motstand [German medicine during the Nazi regime–cooperation and resistance]. Tidsskr Nor Laegeforen. 1995 Dec 10;115(30):3794–9. Norwegian. PMID: 8539755.

profession was rare and presented no significant obstacle to the Nazis' success.[12]

National Socialist German Physicians' League (*Nationalsozialistischer Deutscher Ärztebund*) was founded on August 3, 1929, during the annual Nazi Party Rally in Nuremberg. In 1936, more than 30% of German doctors belonged to the Nazi party.[13] Membership grew incessantly; in the Rhineland, for example, the proportion of NSDAP members among doctors increased from 37% in 1936 to 56% in 1944.[2] By 1945, more than 7% of all German physicians had even become members of the SS, while the figure for the general population was less than 1%.[14]

The German medical profession had thus become a willing instrument that enabled Hitler to implement his devastating policies of race hygiene: involuntary sterilization, the murder of vulnerable patients, and finally genocide. While murder and genocide are self-evidently barbaric, the profound evil of involuntary sterilization, the focus of this book, might not be as immediately obvious.

- It denies individuals control over their own reproductive capacity, a cornerstone of personal freedom.
- It disregards their right to make decisions about their own bodies.
- It violates the fundamental ethical principle of autonomy, equality, dignity, and informed consent.
- It causes permanent harm, including severe psychological and physical trauma.

These are some of the critical issues we will explore in the following chapters.

Box 1: Hitler's Beer Hall Putsch

- This attempt by Adolf Hitler and the Nazi Party to overthrow the Bavarian government took place on November 8–9, 1923, in Munich.
- Its aim was to spark a national revolution against the Weimar Republic.
- Hitler and his supporters stormed a beer hall where a political meeting was held, taking hostages and declaring a national revolution.

---

[12] Ernst E. 50 years ago: the Nuremberg doctors' tribunal. Part 2: Medical resistance during the Third Reich. Wien Med Wochenschr. 1996;146(24):629–31. PMID: 9123951.

[13] Geschichte der Medizin: Ärzte im Nationalsozialismus (aerzteblatt.de).

[14] Colaianni A. A long shadow: Nazi doctors, moral vulnerability and contemporary medical culture. J Med Ethics. 2012 Jul;38(7):435–8.

- The next day, about 2000 men marched through Munich.
- They were confronted by police, who opened fire.
- 16 putschists and 4 policemen were killed.
- Hitler fled but was later arrested.
- He was tried for treason and sentenced to five years in prison.
- He served only 9 months during which time he wrote his book "Mein Kampf."
- The putsch solidified Hitler's leadership within the Nazi Party.

Box 2: Fritz Gottlieb Karl Lenz (1887–1976)

- Born in Pflugrade, Pommern.
- Studied medicine in Berlin and Freiburg.
- 1909, joined the Society for Race Hygiene.
- In Freiburg, he became a pupil of Alfred Ploetz.
- 1913, he became the editor of the "Archiv für Rassen- und Gesellschafts-Biologie" (Archive for Racial and Social Biology).
- 1919, Habilitation (PhD) under Max von Gruber in Munich.
- Lenz co-published the influential book "Grundriss der menschlichen Erblichkeitslehre und Rassenhygiene" (Basic Outline of Human Heredity and Racial Hygiene).
- Hitler incorporated Lenz's ideas into his book "Mein Kampf."
- 1923, Lenz was appointed to the first chair of Racial Hygiene in Germany at the University of Munich.
- In 1931, he called for the "most unfit third of the population" to be sterilized.
- In 1933, Lenz took over the Chair of Social Hygiene at the University of Berlin.
- In the same year, he became Director of the Eugenics Department at the Kaiser Wilhelm Institute of Anthropology in Berlin as well as a member of the "Expert Advisory Council for Population and Racial Policy to the Reich Minister of the Interior" and was involved in the formulation of the "Law for the Prevention of Hereditarily Diseased Offspring," which legalized forced sterilizations.
- In 1937, Lenz joined the NSDAP.
- In 1940, Lenz was involved in the creation of the "euthanasia law" (Hitler refused to pass the law and wanted to wait until the end of the war).
- In 1944, Lenz left Berlin together with his family.
- After the war, Lenz became associate professor and, in 1952, a full professor of "Human Heredity" in Göttingen.
- He retired in 1955 and died in Göttingen aged 89.

2

Involuntary Sterilization

Involuntary sterilization was an essential part of the Nazi ideology of "race hygiene." It was aimed at eliminating genetically "inferior blood" from the German gene pool. Nazi physicians experimented with various methods of sterilizing even larger populations. These were brutal and often deadly but failed to yield a practical solution.

Involuntary Sterilization as a Part of "Race Hygiene"

The Nazis implemented involuntary sterilization to allegedly protect Germany from "inferior blood" and genetic defects.[1] They argued that genetically healthy families were having fewer children than "genetically inferior" citizens. This, they claimed, would inevitably lead to an increase in "sick and asocial" offspring and result in a burden for Germany.[2] Forced sterilization was deemed to be the answer to this ever-increasing problem.

On July 14, 1933, the Nazis passed the "Law for the Prevention of Genetically Diseased Offspring." It allowed for the identification and subsequent

---

[1] Ernst E. Killing in the name of healing: the active role of the German medical profession during the Third Reich. Am J Med. 1996 May;100(5):579–81.

[2] Nazi Persecution of the Mentally & Physically Disabled (jewishvirtuallibrary.org).

E. Ernst, *The Leopard Lily Project*, https://doi.org/10.1007/978-3-032-24771-1_2

sterilization of individuals with supposed genetic defects, including those with:

- genetic blindness and deafness,
- manic depression,
- schizophrenia,
- epilepsy,
- congenital "feeble mindedness,"
- Huntington's chorea,
- alcoholism.[3]

An often forgotten issue was the mixed-race African/German and Vietnamese/German children born in the early 1920s when troops drawn from the French colonial empire occupied the Rhineland. Racial anthropologists denounced them as "Rhineland Bastards," collected details on them, and persuaded the Nazi public health authorities to sterilize 385 of them in 1937.[4]

The sterilization process involved identification by doctors, examination by a panel of experts, and subsequent sterilization by appointed physicians. Approximately 220 hereditary health courts and 30 higher hereditary health courts were established to oversee this process.[1]

Vasectomy was the method of choice for men and tubal ligation for women. An estimated 300,000 to 400,000 German citizens were thus forced to be sterilized. The procedures caused around 6000 deaths, mostly in women.[5] Due to the complexity and risks of these operations, the Nazis initiated a range of research programs to explore alternative methods. Several options were thus tested on camp prisoners at the Auschwitz (Box 1) and Ravensbrück concentration camps (Box 2). Experimental drugs or caustic substances were employed, carbon dioxide was injected, and X-rays were administered in the pursuit of finding a method of sterilizing large groups of victims (see below).[3]

For some leading proponents of "race hygiene," the sterilization law did not go far enough. They argued that it created human "ballast" and an economic burden that ought to be dealt with by other means. The Nazis therefore perverted the concept of euthanasia from voluntary assisted death

---

[3] Sterilization in Nazi Germany (thoughtco.com).

[4] Weindling P. The Dangers of White Supremacy: Nazi Sterilization and Its Mixed-Race Adolescent Victims. Am J Public Health. 2022 Feb;112(2):248–254. https://doi.org/10.2105/AJPH.2021.306593. PMID: 35080945; PMCID: PMC8802574.

[5] Heesch: Zwangssterilisierungen Kranker und Behinderter (akens.org).

to medically supervised murder, i.e., the infamous "Aktion T4." While the sterilization program was aimed at eliminating diseases, the "euthanasia programs" would eliminate the diseased.[6]

Involuntary Sterilization as a Means of Ethnic Cleansing

While vasectomy for men and tubal ligation for women were workable means of sterilizing relatively small groups of people, these methods were impractical for rendering vast populations infertile. Yet, after the Nazis had invaded large parts of Europe and submitted millions to slave labor, achieving such an aim was deemed necessary. In a 1942 speech given at the SS-Junker School in Bad Tölz, Himmler explained the problem as he saw it:

> "After the victory of the German Nation, we still must develop and colonise the Eastern territory and incorporate it into the area of European culture. Within 20 years after the end of the war—a task which I have set for myself and hope to achieve together with you—the German border will be pushed 500 km further to the East. We shall have to send our peasant families there, organise a migration of the best Germanic blood, and bind the many millions of Russian people to our ends. It means that, when the bells across the world announce the greatest victory of our nation, the busiest time of our lives will begin. Twenty years of struggle for the conquest of the world are ahead... Then the East will be free of foreign blood and our families will settle there as landowners."[7]

To render "the East free of foreign blood," the Nazis needed to consider the forced sterilization of these vast populations. The way they understood it, this strategy would have two equally essential advantages:

- It would enable them in the short-term to continue exploiting this huge workforce which, of course, had become an urgent necessity in the pursuit of the war effort.
- In the long-term, it would allow the Nazis to ensure that these "inferior races" would not reproduce and hinder the realization of the Nazi's vision of a continent free of Jews, Slavs, Romani, and other "non-Aryan" races.

---

6 The Nazi Slaughter of the Disabled: The Euthanasia Program T4: Amazon.co.uk: Gerstein, Kurt: 9781937727710: Books Gerstein K: The Nazi Slaughter of the Disabled: The Euthanasia Program T4. American Bibliographical Press (2017).

7 'Three Million Bolsheviks Could Be Sterilised' | Project | Nuremberg. Casus pacis.

Obersturmbannführer-SS Rudolf Brandt, Personal Administrative Officer to Reichsführer-SS Himmler, had outlined the goals of the project with typical bluntness:

> Himmler was extremely interested in the development of a cheap and rapid sterilization method which could be used against enemies of Germany, such as Russians, Poles and Jews. The capacity for work of the sterilized persons could be exploited by Germany, while the danger of propagation would be eliminated.[6]

In an attempt to put this plan into action, the Nazis began testing several methods of mass sterilization. Himmler had been making plans since 1941 to use the Auschwitz concentration camp (Box 1) as a location for such experiments. A proposal had been put forward by Prof. Carl Clauberg (Box 3) to set up a research center for his experiments on a non-surgical method to sterilize women. These experiments thus began in Auschwitz in mid-1942.[8]

Chemical Sterilization

Clauberg's sterilization experiments might be best summed up by the report he himself sent to Himmler on June 7, 1943:

> The method I have invented for the non-surgical sterilization of the female body is nearly ready. It consists of one intrauterine injection and may be administered by practically any physician who can perform a gynaecological examination. I have written that it is 'nearly ready', which means that (1) it only needs some minor adjustments, (2) it can be implemented in our regular eugenic sterilizations already now instead of surgery and can replace surgery… If the results of my research continue along the same line as what I have had up to now (and there is no reason to suppose that they will not), then right now we are not far from the time when one properly trained physician working in an appropriately equipped facility with about ten ancillary staff (this number of ancillary staff corresponds to the required rate of treatment) will probably be able to perform hundreds of sterilizations, if not a thousand, in one day.[7]

Between 150 and 400 Jewish women from various countries became the victims of Clauberg's barbaric experiments. Under the pretext of performing a gynecological examination, Clauberg first checked whether his victim's Fallopian tubes were open. Subsequently, he introduced a chemical irritant, e.g., formaldehyde or silver nitrate, in order to cause a local inflammatory

---

[8] X-ray "sterilisation" and castration in Auschwitz: Dr Horst Schumann | Medical Review Auschwitz

reaction. This, in turn, prompted a permanent obstruction of the tubes followed by infertility. The results were then checked using radiography. The procedures were carried out in a most brutal way, and complications were frequent. They included peritonitis and hemorrhages from the reproductive tract, high fever, sepsis, and multiple organ failure. Many—the exact number is not known—of Clauberg's victims died, and others were later murdered so that autopsies could be carried out to obtain objectively documented results.[9]

Radiation

Dr Horst Schumann, one of the most notorious of all Nazi physicians (Box 4), was in charge of the X-ray irradiation experiments on male and female prisoners in Auschwitz. He irradiated about 30 victims per session conducting two or three such sessions per week. The ovaries or testicles of his victims were submitted to intense radiation. A few weeks later, the victims' organs were surgically removed to verify the changes that occurred after different doses of irradiation.

The treatment left the victims with severe radiation burns and suppurating wounds that often resisted healing.[10] Many prisoners died—exact numbers are not known—some were able to resume work, others committed suicide. Those who were not fit to go back to work were murdered.[7] Only very few of the hundreds, perhaps even thousands of victims survived.[8]

Hormonal Treatment

It has been suggested that SS doctors also experimented with hormonal treatments to induce sterility. However, direct evidence for this assumption is missing. Himmler ordered all participants of the sterilization experiments to keep their investigations secret. Consequently, many experiments, especially those involving novel methods like hormones, are not well-documented.

The collective results of these experiments made it soon obvious that none of the tested methods would lead to a truly practical method of sterilizing vast populations. Chemical methods required a gynecological intervention, irradiation needed X-ray equipment, and hormonal methods were still in their early infancy. For the largest project of ethnic cleansing ever conceived, Himmler would need something different:

- a method that would not impede the victim's ability to work,

---

9 Carl Clauberg/Medical experiments/History/Auschwitz-Birkenau
10 Horst Schumann/Medical experiments/History/Auschwitz-Birkenau

- a treatment that could be given by mouth,
- a method that would work without being noticed by the victim.

As we shall see in the following chapters, such an option seemed to emerge from an entirely unexpected corner: naturopathy.

Box 1: Auschwitz Concentration Camp

- The largest complex of prison camps in Nazi Germany.
- It was in operation from 1940 to 1945.
- It consisted of three successively expanded large concentration camps and around 50 satellite camps.
- It functioned both as a concentration camp and an extermination camp.
- The latter was called Birkenau.
- Various human experiments on prisoners took place in Auschwitz.
- The prisoners came from Belgium, Germany, France, Greece, Italy, Yugoslavia, Luxembourg, the Netherlands, Austria, Poland, Romania, the Soviet Union, Czechoslovakia, and Hungary.
- Around 90% of them were Jews.
- The death toll at Auschwitz was between 1.1 and 1.5 million people.
- The camps were liberated by the Red Army on January 27, 1945.

Box 2: Ravensbrück Concentration Camp

- The camp was established in 1938/1939 in the municipality of Ravensbrück in the province of Brandenburg.
- After the first female prisoners had arrived, they were forced to build the numerous extensions of the original camp.
- Ravensbrück was a complex composed of the neighboring men's camp, industrial plants, the Uckermark concentration camp for girls and young women, the Ravensbrück Siemens industrial camp, as well as a large number of further subcamps.
- About 130,000 women and children, 20,000 men and 1000 young women from 40 nations and ethnic groups were interned in Ravensbrück.
- 28,000 prisoners are thought to have died in the Ravensbrück complex.
- Prisoners had to perform various forms of forced labor mostly aimed at aiding the German war effort.
- From 1942, several further types of medical experiments on prisoners were conducted.
- In February 1945, an execution site and a provisional gas chamber were set up in Ravensbrück, where 2,300 to 2,400 prisoners were killed by the end of March.
- Soviet troops reached the camp on April 30 and liberated the remaining inmates.
- The former main camp served as barracks for the Soviet armed forces from 1945 to 1993.
- The Ravensbrück Memorial was opened in 1959 and later expanded several times.

Box 3: Carl Clauberg[11]

- Born in 1898 in Wupperhof.
- Became SA member in 1933.
- 1937, Professor in Koenigsberg.
- 1942, begin of experiments in Auschwitz (see above).
- 1945, Ravensbrück concentration camp.
- After the war, imprisonment by the Soviets.
- 1948, sentenced to 25 years imprisonment.
- 1955, transferred to West Germany as part of a prisoner exchange program.
- 1955, imprisonment in West Germany.
- 1957, death in prison.

Box 4: Horst Schumann

- Born in Jahr 1906 in Halle an der Saale.
- 1930, joined the NSDAP.
- 1932, joined the SA.
- 1933, graduated in medicine.
- 1934, public health doctor in Halle.
- 1939, employment for the "Aktion T4."[12]
- 1940, head of the Grafeneck euthanasia center.
- Same year ordered to the Sonnenstein euthanasia center.
- 1940, involvement in the "Aktion 13f13."[13]
- 1941, Auschwitz where he conducted his experiment with X-rays (see above).
- In Auschwitz, he also performed experiments injecting prisoners with typhus.
- 1944, appointed to the Sonnenstein Clinic, a military hospital.
- 1945, captured by the Americans but released, as his identity was not yet known.
- 1946, medical practice in West Germany.
- Fled to Egypt and later Sudan after his identity had been revealed.
- 1947, director of a hospital in Sudan.
- 1962, fled to Ghana after being recognized by an Auschwitz survivor.
- 1966, extradited from Ghana to West Germany.
- 1970, put on trial in Frankfurt.
- He admitted to killing 80,000 Jews.
- 1972, released from prison due to his deteriorating health.
- Schumann died in 1983.

---

[11] Carl ClaubergWikipedia.

[12] Aktion T4Wikipedia.

[13] Action 14f13—Wikipedia.

3

Naturopathy and the "Neue Deutsche Heilkunde"

In the early-twentieth century, various naturopathic practices were popular in Germany. The Nazis sought to address the tension between naturopathic practitioners and conventional doctors by creating the "Heilpraktiker" and the "New German Medicine" along the lines of their ideology.

Naturopathy is the medical treatment of illness with the means supplied by nature. It includes many modalities that were popular in the early parts of the twentieth century, e.g.:

- cold, e.g., ice packs,
- dietary approaches,
- energy healing,
- exercise therapy,
- heat, e.g., mud packs,
- herbal medicine,
- homeopathy,
- hydrotherapy
- iridology,
- massage therapy,
- radionics,
- Schüssler salts.

E. Ernst, *The Leopard Lily Project*, https://doi.org/10.1007/978-3-032-24771-1_3

Naturopathy was close to the heart of many leading Nazis, particularly Himmler (Chap. 5). Driven by a general fascination with nature, Himmler felt that naturopathy was aligning perfectly with the ideology of "race hygiene" (Chap. 1), the Nazi culture of "Blut und Boden" (blood and soil)[1] and with Germanic traditions.

Yet, tensions persisted between the many different types of naturopathic lay healers on the one hand and conventional doctors on the other. The medical establishment remained skeptical about the value of many of the treatments in question, and professional disagreements arose regularly, for instance, over practices like vaccination, which lay healers tended to oppose and conventional physicians strongly supported. The Nazis' approach of solving this problem was the creation of the "Neue Deutsche Heilkunde" (New German Medicine, NGM).

The powerful nature-orientated movement that had existed long before the "Third Reich" was driven not least by consumers who supported it and were organized in associations like the "Kneippbund," the "Priessnitzbund," the "Reichsbund für Homöopathie," and the "Biochemischer Bund," all of which represented specific alternative therapies. By the early 1930s, these associations had about half a million members and an estimated further 5–10 million supporters. When the Nazis seized power in 1933, the number of non-medically qualified, non-licensed lay healers was roughly equal to that of licensed physicians.[2]

The Nazis thought of solving this tension with a cunning, three-pronged strategy aimed at satisfying all parties.[2]

- By unifying all the highly diverse branches of lay healers and giving them the official title of "Heilpraktiker," they wanted to satisfy the healers' desire for recognition and status.
- Since the "Heilpraktiker" would become extinct within one generation, the concerns of conventional physicians were taken care of.
- In parallel, conventional doctors would be converted to physicians practicing the NGM which would satisfy the many leading Nazis who align conventional medicine with their ideology.

---

[1] Blood and soil—Wikipedia.

[2] Ernst E. 'Neue Deutsche Heilkunde': complementary/alternative medicine in the Third Reich. Complement Ther Med. 2001 Mar;9(1):49–51. https://doi.org/10.1054/ctim.2000.0416. PMID: 11,264,971.

Dr Gerhard Wagner, the Reich's chief medical officer (Box 1), was an influential driving force behind this strategy. He wrote in 1935: "If today we want to construct a NGM, then its basis cannot be that formed by the exact sciences, but its basis must be our national socialist world view."[3]

The "Heilpraktikergesetz" (law regulating the Heilpraktiker) was passed on February 17, 1939, and went into effect on February 21, 1939. It was the legislation designed to regulate and, ultimately, phase out the profession of "Heilpraktiker."

- It established that only licensed individuals could practice medicine without a formal medical degree.
- It decreed that the use of the title "Heilpraktiker" was restricted to those who had been previously practicing.
- It regulated that no further "Heilpraktiker" could be educated.
- It criminalized the unlicensed practice of medicine, with penalties including imprisonment and fines.

The law was a pivotal part of the Nazi regime's broader effort to control and consolidate all aspects of German life, including medicine, under a unified, state-controlled system. While it seemingly established and regulated the new profession, its true purpose was to eliminate lay healers within just a few decades.

In the profession's journal "Der Heilpraktiker," the structure and task of the profession's organization, the "Heilpraktikerbund," were outlined as follows:

In accordance with the Führer principle, the entire initiative in the Heilpraktikerbund Deutschlands comes from its federal leader, party comrade Ernst Kees. All employees are therefore primarily executive organs of the federal leader …. The federal leader was appointed by the Reich Minister of the Interior at the end of March 1934 on the recommendation of the Deputy of the Führer. He was given the task by the government and the state of purging the

---

3 Wagner G. Volksgesundheitswacht, Zeitschrift des Sachverständigenbeirates für Volksgesundheit bei der Reichsleitung der NSDAP. München Nr. 12, 1935.

Heilpraktikerbund of all useless and unreliable elements that appeared unacceptable to the new state and whose elimination was in the interests of public health ....[4]

The "Heilpraktiker" had rights, duties and standing like those of physicians—except for one crucial point: They were not permitted to open schools and educate a next generation of "Heilpraktiker." Therefore, they were destined to become extinct within merely one generation. Goebbels called the law the "cradle and the grave" of a new profession.[5]

An essential aspect of the NGM was, of course, that of plant-based medicine. Nazi ideology viewed this not merely natural but also economical and intrinsically German. Himmler thus initiated several research projects in this area. One example that sands out for its cruelty and scientific uselessness is the Polygal project of 1942–1944 which we will discuss in more detail in Chap 5.

The creation of the NGM as well as the "Heilpraktiker" had far-reaching consequences and impacts on German health care to the present day (Box 2). It was to a large extent driven by the enthusiasm for alternative medicine shared by key figures within the Nazi regime, a subject that we shall explore in more detail in the next chapter.

Box 1: Gerhard Wagner (1888–1939)[6]

- Gerhard Wagner studied medicine and served as a medical officer during World War I.
- He then practiced medicine in Munich where he joined the NSDAP in 1929.
- He established the "National Socialist German Doctors' League" in 1932 and became its leader.
- In 1934, he was appointed "Reichsärzteführer" (Reich Doctors' Leader), thus becoming the chief medical officer of Germany.
- He shaped Nazi health policies, including the Nuremberg Race Laws (Chap. 1).
- He was a strong proponent of eugenics and supported policies such as the sterilization of individuals deemed "unworthy of life" (Chap. 2).
- Gerhard Wagner died of cancer in Munich in March 1939.
- He was succeeded by Leonardo Conti.[7]

---

[4] Heilpraktiker

[5] Donner F. Bemerkungen zu-der Überprüfung der Homöopathie durch das Reichsgesundheitsamt 1936–1939. Perfusion 1995; 8: 3–7 (part 1), 35–40 (part 2), 84–88 (part 3), 124–129 (part 4), 164–166 (part 5).

[6] Gerhard Wagner (physician)—Wikipedia.

[7] Leonardo ContiWikipedia.

Box 2: The German "Heilpraktiker" Today[8]

- In 1957, the "Heilpraktiker" profession legally challenged and managed to overturn the prohibition to educate further practitioners.
- Therefore, a large group of these practitioners is currently allowed to practice.
- "Heilpraktiker" do not have to attend any formal education or training.
- Once they have passed their exam, they are free to practice, even if they have no clinical experience whatsoever.
- This led to a two-tier health system with regularly trained and educated physicians on the one hand and "Heilpraktiker" on the other side.
- Many experts have warned that this situation puts people's health at risk.
- So far, a decisive political will to change it has not emerged.

---

[8] Vorsicht Heilpraktiker: Eine kritische Analyse: Ernst, Edzard: Amazon.de: Books Ernst E: Vorsicht Heilpraktiker; Eine Kritische Analyse. Springer (2023).

4

Top Nazis and Alternative Medicine

During the "Third Reich," alternative medicine was promoted by several leading Nazis who saw naturopathy and "natural" lifestyles as aligning with their ideology of racial purity and Germanic traditions. After the war, many of the physicians who had promoted these ideas were able to continue their careers, and some even received prestigious awards.

The extraordinary push toward alternative medicine during the "Third Reich" was due not least to the fact that several Nazis in leading positions were proponents of this form of health care. Here are a few prominent examples of Nazi officials supporting the "New German Medicine," NGM (Chap. 3):

Rudolf Hess (1894–1987) was appointed as Hitler's deputy on April 21, 1933. The two men had long been close friends and had both served a prison sentence in Landsberg after the failed "Beer Hall Putsch" where Hess had been invaluable assistance for Hitler writing his book "Mein Kampf" (Chap. 1).

Hess became obsessed with his health and regularly consulted medical and non-medical practitioners for his imagined ailments which included ailments of the kidneys, colon, gall bladder, bowels, and heart. He belonged to the "Thule Gesellschaft"[1] (Box 1), was a vegetarian, did neither smoke nor drink,

---

[1] Thule Society—Wikipedia.

E. Ernst, *The Leopard Lily Project*, https://doi.org/10.1007/978-3-032-24771-1_4

and had a keen interest in astrology, clairvoyance, the occult as well as alternative medicine.[2] A fitting illustration for his enthusiasm is his speech as the patron of the 1937 World Conference on Homeopathy in Berlin:

> "It is known that not just novel therapies but also traditional ones, such as homeopathy, suffer opposition and rejection by some doctors without having ever been subjected to serious tests. The doctor is in charge of medical treatment; he is thus responsible foremost for making sure all knowledge and all methods are employed for the benefit of public health…I ask the medical profession to consider even previously excluded therapies with an open mind. It is necessary that an unbiased evaluation takes place, not just of the theories but also of the clinical effectiveness of alternative medicine.
>
> More often than once has science, when it relied on theory alone, arrived at verdicts which later had to be overturned—frequently this occurred only after long periods of time, after progress had been hindered and most acclaimed pioneers had suffered serious injustice… Insightful doctors, some of whom famous, have, during the recent years, spoken openly about the crisis in medicine and the dead end that health care has maneuverer itself into. It seems obvious that the solution is going in directions which embrace nature. Hardly any other form of science is so tightly bound to nature as is the science occupied with healing living creatures. The demand for holism is getting stronger and stronger, a general demand which has already been fruitful on the political level. For medicine, the challenge is to treat more than previously by influencing the whole organism when we aim to heal a diseased organ."

The "Rudolf Hess Krankenhaus" in Dresden was a project dedicated to the integration of conventional and alternative medicine. It was meant to be the model after which most, if not all, German medicine was to be conceived. The treatments offered consisted of natural and "Aryan" methods with an emphasis on homeopathy, herbalism, hydrotherapy, and fasting. Two unforeseen events significantly limited the long-term impact of the experiment. Firstly, the Reich's chief medical officer and ardent supporter of the NGM, Dr Gerhard Wagner (see below), died on March 25, 1939. He was replaced by Dr Leonardo Conti who showed much less interest in alternative medicine. Secondly, on May 10, 1941, Hess secretly flew to Scotland and was declared a mentally deranged traitor (Box 2). Subsequently, the institution was renamed "Gerhard-Wagner Krankenhaus" and functioned henceforth as a regular hospital.[3]

---

[2] Rudolf Hess: Der Stellvertreter (German Edition) eBook: Görtemaker, Manfred: Amazon.co.uk: Books Görtemaker M: Rudolf Hess: Der Stellvertreter. C.H.Beck (2023).

[3] Otto E. Das Dresdner Experiment: Naturheilmethoden sollten überprüft werden. Dtsch Ärzteblatt 1993; 90: 948–951.

Hess' enthusiasm for alternative medicine is further exemplified by his support of the German Ministry of Health's research project on homeopathy. The project became (and still is) the most comprehensive investigation in the history of homeopathy. It scrutinized homeopathy and its assumptions through systematic research supervised by some of the best conventional physicians and scientists of the period. About 60 university institutions were involved; each research team included homeopaths, toxicologists, pharmacologists, and internists. Around 300 planning meetings with staff from the ministry took place where large, placebo-controlled trials were discussed on patients with tuberculosis, pernicious anemia, gonorrhea, and other diseases for which homeopaths had previously claimed particularly convincing success. An eyewitness, Dr. Fritz Donner, described several clinical trials in some detail and reported that, without exception, their results failed to support the claims made by homeopaths. Due to the outbreak of WWII, the project was aborted prematurely, and its full results were never published.[4]

Julius Streicher (1885–1946) was the editor of the virulently anti-semitic newspaper, "Der Stürmer" (The Attacker), and a participant in Hitler's "Beer Hall Putsch" of 1923 (Chap. 1, Box 1). After the Nazis had seized power, in 1933, Streicher chaired the "Zentralkomitee zur Abwehr der jüdischen Greuel- und Boykotthetze" ("Central Committee to Repulse Jewish Atrocity and Boycott Agitation") and helped organize the nationwide boycott of Jewish businesses on April 1, 1933.[5]

Like Hess, Streicher was a member of the "Thule Gesellschaft" (Box 1) and an enthusiastic supporter of alternative medicine. He published a pseudomedical Journal entitled "Deutsche Volksgesundheit aus Blut und Boden!" (German Public Health through Blood and Soil!). Even for Nazi standards, the publication went too far claiming that vaccination was part of the Jewish plot against the Aryan race. Thus it was discontinued in 1935. In 1937, Streicher made sure that one of Germany's most prominent medical proponents of alternative medicine, Karl Kötschau (see below),[6] was appointed director of a prestigious university department in Nuremberg.

After the war, Streicher was convicted at the Nuremberg Trials for crimes against humanity. The tribunal noted that his propaganda contributed to the

---

4 Donner F. Bemerkungen zu-der Überprüfung der Homöopathie durch das Reichsgesundheitsamt 1936–1939. Perfusion 1995; 8: 3–7 (part 1), 35–40 (part 2), 84–88 (part 3), 124–129 (part 4), 164–166 (part 5).
5 Julius StreicherWikipedia.
6 Deutsche Biographie—Koetschau, Karl.

psychological preparation for the Holocaust. He was executed by hanging on October 16, 1946.

Georg Gustav Wegener (1895–?) was a member of the SS and a key proponent of the NDH (Chap. 3). In 1938, he became the director of the "Reichsarbeitsgemeinschaft fuer Heilpflanzenkunde und -beschaffung" ("Reich Working Group for Herbal Medicine and Acquisition"). In 1935, he was appointed the leader of the newly formed "Reichsarbeitsgemeinschaft der Verbände fuer naturgemässe Lebens- und Heilweise" (Reich Working Group of Associations for Natural Ways of Life and Healing). The task of this organization was to bring together all the multiple groups of lay healers under one single umbrella. It was dissolved in 1941 and replaced by the "Deutschen Volksgesundheitsbund" ("German People's Health Association") which from then on functioned as the central organization for the promotion of public health.

In 1938, together with two prominent co-authors, Wegener published the book "Allgemeine Heilpflanzenkunde Grundlagen einer rationellen Gewinnung, Verarbeitung, Anwendung und Erforschung der Heil- und Gewürzpflanzen" (General medicinal plant science and fundamentals of rational harvesting, processing, application and research of medicinal and aromatic plants).[7] In 1941, he became the head-investigator in the herb plantation of the Dachau concentration camp (Chap. 3). Wegener's fate after the war is unknown.[8]

Joseph Goebbels (1894–1945), the Minister of Propaganda, was a key official in publicizing the Nazi health messages. He used propaganda to promote a "natural" and "wholesome" German lifestyle, a core component of the NGM. He saw these ideas as a way to reinforce racial purity and national health and employed his media control to attack conventional medicine which he defamed as Jewish-dominated. Goebbels was deeply involved in the atrocities committed by the Nazis. He killed himself and his family hours after Hitler had committed suicide in his Berlin bunker.

Baldur von Schirach (1907–1974), the leader of the "Hitler Youth,"[9] was deeply involved in promoting physical fitness and "natural" lifestyles among German youths. He promoted hiking, gymnastics, and outdoor activities, which were all seen as ways to build a strong and healthy "Volkskörper" (body of the people). This focus on physical health and a "natural" lifestyle was

---

[7] Ernst Günther Schenck, Rudolf Lucass, G G. Wegener: Allgemeine Heilpflanzenkunde Grundlagen einer rationellen Gewinnung, Verarbeitung, Anwendung und Erforschung der Heil- und Gewürzpflanzen. Wilhelm Heyne Verlag, Dresden, 1938.

[8] Klee E: Das Personenlexikon zum Dritten Reich. Fischer, Frankfurt, 2015.

[9] Hitler YouthWikipedia.

meant to complement the concept of the NGM. In this role, Schirach was responsible for brain-washing millions of young Germans with Nazi propaganda. After stepping down from this role, he was appointed "Gauleiter" (district leader) and "Reichsstatthalter" (Reich governor) of Vienna. After the war, Schirach was put on trial at the Nuremberg Tribunals and convicted of crimes against humanity for his part in the deportation of over 65,000 Jews from Vienna. He was sentenced to 20 years in prison, served his full sentence, and was released in 1966. He died in 1974.

The most ardent supporters of the idea of the NGM were prominent members of the German medical profession.

Alfred Brauchle (1898–1964) joined the NSDAP early in his career and became the medical leader of the NGM movement. He was appointed director of the department of naturopathy at the Rudolf Hess Krankenhaus and was instrumental in aligning the German health care with the Nazi ideology. In 1942, Hitler made him a professor. After the war, he was able to return to work and significantly shaped post-war alternative medicine. For his achievements in alternative medicine, the "Association of Physicians for Naturopathy and Regulatory Medicine" awarded him the Hufeland Medal in 1958, the highest award in this field.[10]

Sigfried Gräff (1887–1966) became a member of numerous NS organizations: the Stahlhelm, the National Socialist People's Welfare Organization, the Reich Association of Large Families, the Reich Association of German Civil Servants, the NSDAP, the National Socialist German Doctors' Association, National Socialist Teachers' Association, and the Reich Colonial Association. In 1939, Gräff was awarded the title of "associate professor." Gräff wore a uniform during his public lectures. After the war, he continued his work unhindered. In 1959, he received the Hufeland Medal.[11]

Hans Haferkamp (1906–1982) served as camp doctor at the Dachau concentration camp between 1942 and 1944. Together with the infamous Sigmund Rascher,[12] he published an article about their experiments on the natural remedy Polygal in the Muenchner Medizinische Wochenschrift in 1944. It was based on unspeakably cruel experiments on camp prisoners which resulted in the death of several victims (Chap. 5). After the war, he was able to continue practicing as a naturopathic doctor. In 1951, he became the director of postgraduate education for the "Association of Physicians for

---

[10] Alfred BrauchleWikipedia.
[11] Siegfried Gräff—Wikipedia.
[12] Three Meetings with the Nazi Doctor Sigmund Rascher | Skeptical Inquirer Sept/Oct 2025.

Naturopathy and Regulatory Medicine" and received their Hufeland Medal in 1975.[13]

Wolfgang Kohlrausch (1888–1980) was appointed associate professor at the University of Berlin in 1934. From 1935 to 1941, he was head of the Sports Medicine Institute at the Medical Faculty of the University of Freiburg. In 1937, he joined the NSDAP and was appointed "Chief Medical Officer" of the "Hitler Youth." He also was a member of the National Socialist Motor Corps, the National Socialist German Doctors' Association, the NS-Altherrenschaft, and the National Socialist German Lecturers' Association. From 1941 to 1944, he was a full professor of movement therapy at the Reich University of Strasbourg. After the war, he was imprisoned for one year. Subsequently he was able to resume his university career on a part-time basis. In 1966, he was awarded the Hufeland Medal.[14]

Werner Kollath (1892–1970) was an early member of the NSDAP and a supporting member of the SS. He also belonged to the National Socialist Teachers' League, the National Socialist Lecturers' League, the National Socialist People's Welfare Organization, and the Reich Air Defense League. In 1935, he was appointed Professor of Hygiene and Bacteriology at the University of Rostock and deputy medical assessor at the Rostock Hereditary Health Court. His textbooks promoted both racial hygiene and whole food nutrition. After the war, he worked as a food chemist and consultant for the biscuit manufacturer Bahlsen in Hannover.[15] He was awarded the Hufeland Medal in 1957.[16]

Karl Kötschau (1892–1982) replaced a Jewish professor at the University of Jena in 1934 and gave lectures promoting "racial hygiene" and "Aryan" healing. His inaugural lecture had the title "The national socialist idea in biological medicine." He was a member of numerous NS organizations and endorsed forced sterilization as well as murder of disabled patients (euphemistically called "euthanasia" by the Nazis) in the name of "race hygiene." After the war, he continued his medical career and received several awards, including the Hufeland Medal in 1968.[17]

Ernst Günter Schenk (1904–1998) was a member of the SA and the NSDAP as well as numerous further NS organizations. In 1940 he became the Inspector for Nutrition of the Waffen-SS and conducted dietary experiments on camp prisoners. In the last days of the "Third Reich," Schenk met

---

[13] Klee E. Das Personenlexikon zum Dritten Reich. Fischer Verlag, Frankfurt, 2015.
[14] Wolfgang Kohlrausch Wikipedia.
[15] Werner Kollath Wikipedia.
[16] Werner Kollath Wikipedia.
[17] Karl Kötschau Wikipedia.

Hitler in person. After the war, he served a 10-year prison sentence under the Soviets.[18] He then worked for a pharmaceutical firm and published a book entitled "Patient Hitler."[19]

Gerhard Wagner (1888–1939) had become a member of the Nazi party as early as 1929 and later became Hitler's "Reichsärzteführer." Until his death from cancer in 1939, he was the most important physician of the "Third Reich." He was a crucial driving force behind many of the Nazi's ideas related to health care, not least the NGM.[20]

By far the most important advocate of alternative medicine during the Third Reich was Heinrich Himmler. He will be the subject of the following chapter.

Box 1: The "Thule Gesellschaft"

- The society was rooted in a mix of occultism and "völkisch" nationalism. Its members subscribed to a mythology of an ancient Aryan homeland, Ultima Thule, and believed in the racial superiority of the Aryan race.
- A central tenet was a deep-seated hatred for Jews and a strong opposition to communism. Members were required to sign a "blood declaration of faith" affirming their Aryan lineage.
- Members were involved in paramilitary activities and counter-revolutionary efforts.
- The society supported the Deutsche Arbeiterpartei (DAP, German Workers' Party) in 1919. It later merged with the NSDAP.
- Several individuals who would later become prominent Nazis were members.
- The society acquired a weekly paper, the "Münchner Beobachter," which was later renamed the Völkischer Beobachter and became the primary Nazi newspaper.
- In the mid-1920s, the Thule Society's influence waned, its founder, Rudolf von Sebottendorff, was arrested and it dissolved.

Box 2: Hess's 1941 Clandestine Flight to Scotland

- Hess flew a Messerschmitt Bf 110 alone from Germany with the goal of negotiating peace with Britain.
- Experts speculate that Hess saw this as his chance to become a hero by securing a separate peace with Britain and allowing Germany to focus its full military might on fighting the Soviet Union.

---

[18] Ernst-Günther SchenckWikipedia.

[19] Patient Hitlerbk1311: Ernst Günther Schenck: Amazon.co.uk: Books Schenck G: Patient Hitler. Weltbild Verlag (2000).

[20] Gerhard Wagner (physician)—Wikipedia.

- When Hitler was informed of the flight, he was outraged and declared Hess to be a traitor and mentally ill stripping him of all his titles and positions.
- The British government viewed Hess's "peace offer" as a trap and held him as a prisoner of war.
- After the war, Hess was accused at the Nuremberg Trials alongside other prominent Nazi leaders.
- The court convicted him of two charges: "conspiracy to commit crimes against peace" and "crimes against peace."
- Hess was sentenced to life in prison. He was the sole prisoner in Spandau prison from 1966 until his death.
- Hess died by suicide on August 17, 1987, aged 93.

5

Heinrich Himmler's Enthusiasm for Alternative Medicine

Heinrich Himmler was fascinated by alternative medicine. His obsession prompted him to authorize several physicians to conduct cruel experiments on concentration camp inmates. They were steeped in pseudoscience and resulted in many deaths. Upon being apprehended in 1945, Himmler committed suicide.

Many excellent books have been written about Heinrich Himmler.[1] Therefore, there is no need to elaborate on his life; instead, I will mention some essential background and focus on Himmler's enthusiasm for all things esoteric, particularly alternative medicine.

Himmler joined the NSDAP in 1925, rose steadily in the party hierarchy, and was elected a deputy to the Reichstag in 1930. He soon became head of the SS,[2] the police, as well as the Gestapo and gained responsibilities in the areas of security, intelligence gathering, and espionage. After the invasion of the Soviet Union, he was also entrusted with the administration of the conquered territories and became head of the German anti-partisan campaign which involved mass murderand numerous atrocities, including the massacre

---

[1] Heinrich Himmler: Amazon.co.uk: Longerich, Peter: 8601420044999: Books Longerich P: Heinrich Himmler. OUP Oxford (2012).

[2] Schutzstaffel—Wikipedia.

E. Ernst, *The Leopard Lily Project*, https://doi.org/10.1007/978-3-032-24771-1_5

of Jews. He became a key architect of the Holocaust, organizing the extermination campsin occupied Poland and taking responsibility for the slaughter of millions.

Raised in a conservative Catholic family in Bavaria, Himmler had developed an early interest in mysticism, influenced by romanticized views of Germanic history and folklore. His interests in this area were numerous and bizarre:,[3,4,5,6]

- Fascinated with ancient Germanic and Nordic myths, he believed they represented the pure Aryan heritage, promoted the worship of Germanic gods like Wotan (Odin), and tried to revive pagan rituals.
- He embraced ideas of Aryan racial superiority and believed Aryans were descendants of a divine or cosmic race linked to pseudoscientific theories about Atlantis or Hyperborea.
- He was fascinated by runic alphabets which he understood as sacred symbols of Aryan wisdom. The SS thus adopted the double Sig rune as its emblem.
- He consulted astrologers and was intrigued by cosmic theories, such as the "World Ice Theory" (Welteislehre)[7] and believed that the universe was shaped by ice and fire.
- He was captivated by the sage of the Holy Grail and sponsored expeditions in search for Grail-related artifacts.
- He transformed Wewelsburg Castle into a "Nordic Vatican" and used the castle's crypt for ceremonial purposes.
- He envisioned the SS as a mystical knightly order.

---

[3] The Occult Roots of Nazism: Secret Aryan Cults and Their Influence on Nazi Ideology: Amazon.co.uk: Goodrick-Clarke, Nicholas: 9781838601850: Books Goodrick-Clarke N: The Occult Roots of Nazism: Secret Aryan Cults and Their Influence on Nazi Ideology. Tauris Parke (2019).

[4] The Nazis and the Occult: The Third Reich's Search for Supernatural Powers: Amazon.co.uk: Roland, Paul: 9781398809239: Books Roland P: The Nazis and the Occult: The Third Reich's Search for Supernatural Powers Sirius Entertainment (2021).

[5] Himmler's Crusade: The Nazi Expedition to Find the Origins of the Aryan Race: Amazon.co.uk: Hale, Christopher: 9780471262923: Books Hale Ch: Himmler's Crusade: The Nazi Expedition to Find the Origins of the Aryan Race. Wiley (2003).

[6] Wewelsburg—Decrypted: Exposing Nazi occultism, the Thule Society and the secret doctrines of Himmler's SS sun-temple.: Amazon.co.uk: Klovekorn PhDCan, Henning A.: 9,798,846,618,923: Books Klovekorn HA: Wewelsburg - Decrypted: Exposing Nazi occultism, the Thule Society and the secret doctrines of Himmler's SS sun-temple. Independently published (2022).

[7] Welteislehre—Wikipedia.

- He funded archeological and anthropological expeditions through the Ahnenerbe[8] hoping to uncover evidence of Aryan supremacy and traces of ancient Aryan civilizations in Tibet, Iceland, and other regions.
- He initiated research into historical witch trials, believing witches were misunderstood bearers of ancient Aryan knowledge persecuted by the Church.
- He framed the Holocaust as a form of "healing" the German race by eliminating perceived racial threats.
- He endorsed alternative healing rituals viewing them as ancient Germanic wisdom and believing they would enhance the spiritual and physical purity of the German nation.

Himmler was also obsessed with his own health, and he saw himself as a scientist with a particularly keen interest in medicine, especially alternative health care. This fascination was rooted in a bizarre blend of mysticism, occultism, pseudoscience, "Aryan" purity, and Germanic heritage. He was besotted with herbal medicine, homeopathy, and natural Germanic traditions. His wife, Margarete, once ran a clinic in Berlin focused on homeopathy, hypnosis, and herbal treatments (Box 1).

The Schüssler Salt Experiments

Schüssler Salts are highly diluted and ineffective remedies akin to homeopathics but based solely on minerals.[9] In 1942, Himmler ordered that Schüssler Salts (Box 2) should be tested on concentration camp prisoners. Great importance was attached to these experiments, so much so that even Ernst Grawitz (Chap. 14) himself made a visit to Dachau to inform himself on progress. For the experiments, 40 Polish priests were injected with pus. Half of them were then treated with sulfonamides (an early version of an antibiotic) and the other half with Schüssler Salts. A third group of completely untreated subjects served as controls. The results seemed to

---

[8] AhnenerbeWikipedia.

[9] Bizarre Medical Ideas: … and the Strange Men Who Invented Them: Amazon.co.uk: Ernst, Edzard: 9783031551017: Books Ernst E: Bizarre Medical Ideas: … and the Strange Men Who Invented Them. Springer (2024).

demonstrate the effectiveness of sulfonamides and confirm the ineffectiveness of Schüssler Salts. All of the victims suffered unspeakable agony, and 28 subjects died.[10]

Several physicians were involved in these experiments:

- Dr. **Heinrich Schütz** (1906–1986) together with Kießwetter (see below) was in charge of the experiments described above. After the war, he was detained by the Allies, released in 1947, and then settled in Essen where he practiced as a specialist in internal medicine. In 1971, he was remanded in custody for the Dachau experiments, and in 1975, he was sentenced to ten years' imprisonment for "accessory to murder and attempted murder" on 11 counts. Due to a medically certified incapacity, Schütz did not have to serve his prison sentence. He died in 1986 in Feldafing, Germany.
- The Austrian camp doctor, **Karl Babor** (1918–1964), personally assisted the experiments. Babor had been a camp doctor at the Großrosen concentration camp where he had killed typhus patients using phenol and prussic acid injections. He was awarded the War Cross of Merit 2nd Class for "services rendered in the fight against the typhus epidemic." From mid-June 1942, he served in the Dachau and later in several other concentration camps. After the war, he became a prisoner of war in France. In the early 1950s, Babor fled to Ethiopia and opened a private practice in Addis Abeba. When he was about to get caught, he committed suicide.[11]
- Dr. Waldemar Wolter (1908–1947) served as a further assistant for the Schüssler Salt experiments. Wolter had served as the camp doctor in the Hinzert SS special camp and the Sachsenhausen concentration camp. In 1942, he worked at the Dachau concentration camp. After the war, Wolter was accused of ordering the gassing of 1,400 to 2,700 prisoners. In 1946, Wolter was sentenced to death and executed on May 28, 1947.
- Dr **Rudolf Kießwetter** (1901–1992), an expert in the treatment with Schüssler Salts, injected the pus into the thighs of the Dachau prisoners.[12] Despite the negative results of the treatment, Kießwetter continued to insist on the efficacy of the Schüssler Salt "ferrum phosphoricum." After the war Kießwetter was indicted but managed to escape punishment and continued his medical practice unhindered in East Germany until his death in 1992.

---

[10] Rau P, Voggenreiter M, Ude-Koeller S, Leven K-H: Medizintäter. Boehlau Verlag, Wien. 2022.

[11] Entmenschlichte Medizin: Österreichs Ärzte im Nationalsozialismus: Amazon.co.uk: Ernst, Edzard: 9783662716144: Books Ernst E: Entmenschlichte Medizin: Österreichs Ärzte im Nationalsozialismus. Springer (2026).

[12] Der Nazi-Arzt aus Burg.

The Polygal Experiments

The physician who conducted many of the worst experiments on camp prisoners for Himmler was undoubtedly the infamous Sigmund Rascher (Box 3). A relationship to Karoline Diehl, Himmler's former secretary, had gained him direct access to Himmler. The Reichsfuehrer then allowed him to conduct research in the Dachau concentration camp focusing on topics that were of relevance to the war effort, such as the notorious freezing and low-pressure experiments.[13]

Less well-known is Rascher's investigation on a pectin-based preparation, Polygal, which was claimed to promote blood clotting. Rascher predicted that the prophylactic use of Polygal would reduce bleeding from wounds sustained in combat or during surgery. In his experiments, he simulated combat injuries by having prisoners shot or by amputating their extremities (without anesthesia). Rascher published an article about his clinical experience with Polygal, without specifying the nature of the trials; it concluded, "The tests of this medicine 'Polygal 10' showed no failures under the most varied circumstances." Subsequently, Rascher even founded a company to manufacture Polygal and used prisoners to work in the factory.[14]

Rascher was not merely an over-ambitious, yet mediocre physician turned sadistic slaughterer of innocent prisoners, he also was a serial falsifier of research data. Before the end of the Third Reich, he lost the support of Himmler and was imprisoned for a string of crimes and executed on the orders of Himmler.[15]

Himmler's interest in medicine had been ignited by constant concern for his own fragile health. Later it became increasingly driven by his ideological fanaticism. His support for brutal experiments, like those at Dachau, and his promotion of alternative medicine were merely tools for advancing Nazi ideological goals. This is particularly true for his interest in herbal medicine as a means of mass sterilization, a topic that we will discuss in the following chapters.

---

[13] Medical Experiments performed at Dachau by Dr. Sigmund Rascher for the Luftwaffe.

[14] Berger RL. Nazi science–the Dachau hypothermia experiments. N Engl J Med. 1990 May 17;322(20):1435–40. https://doi.org/10.1056/NEJM199005173222006. PMID: 2184357.

[15] Three Meetings with the Nazi Doctor Sigmund Rascher | Skeptical Inquirer Sept/Oct 2025.

Box 1: Margarete Himmler's Clinic[16]

- Margarete Himmler, née Boden, had worked as a nurse during World War I.
- During the 1920s, she became the co-owner of a private clinic in Berlin.
- She distrusted conventional medicine and was in favor of alternative approaches such as homeopathy, hypnosis, and herbalism.
- In 1928, Margarete married Heinrich and subsequently her involvement in the clinic diminished.
- In the same year, she joined the NSDAP.
- Her anti-semitism emerged in a 1928 letter criticizing a Jewish co-owner of the clinic, Bernhard Hauschildt.

Box 2: Schüssler Salts

- Wilhelm Schüssler (1821–1898) was born in North Germany, started his professional life as a teacher, later studied medicine, obtained his medical license in 1859, and then practiced as a homeopath.
- He became increasingly disenchanted with homeopathy and created his own form of treatment, the "Schüssler Salts" in the mid-1870s.
- Like homeopathic remedies, they are highly diluted but do not rely on homeopathy's axiom of "like cures like."
- Schüssler Salts are biologically implausible.
- To the present day, there is no proof that they are effective.
- Schüssler Salts were largely ignored by conventional medicine but strongly opposed by homeopaths.
- Yet, they did gain a considerable following among lay people.
- After Schüssler's death, several loyal followers made sure his legacy would not be forgotten.
- During the Third Reich, they made a significant comeback and have remained popular in Germany ever since.

Box 3: Sigmund Rascher (1909–1945)

- Studied medicine in Munich.
- Was influenced by anthroposophical medicine.
- Began his career as a pseudoscientific data falsifier with his doctoral thesis on an anthroposophical diagnostic method.
- Sought out Himmler's acquaintance.
- Suggested several pseudoscientific experiments to Himmler.

---

[16] Margarete Himmle—Wikipedia.

- With Himmler's support, he conducted such experiments at the Dachau concentration camp.
- More than 150 prisoners died as a result.
- Rascher fell out of favor, partly due to unrelated fraud, and was imprisoned in various concentration camps.
- A few days before the liberation of the Dachau concentration camp, he was shot in Dachau on Himmler's orders.

6

Herbal Medicine

The Nazis promoted herbal medicine as a core component of their "New German Medicine." This led to significant research into medicinal plants. In 1938, the herb plantation at the Dachau concentration camp was established. Prisoners were forced into hard labor to cultivate medicinal plants. The aim was to make Germany self-sufficient in pharmaceuticals.

Like many other countries, Germany has a long tradition of using plants for the treatment of illness.[1] After the Nazis had taken power in 1933, the push for herbal medicine became stronger:

- Herbal medicine was in line with the Nazi ideology of "Blut und Boden."
- It fitted perfectly into the concepts of "Heilpraktiker" and "NGM."
- It promised to provide an inexpensive type of health care.
- It relied on German tradition and supplies.

As part of the NGM, the Nazi regime thus initiated significant research into herbal medicine. The aim was to integrate traditional Germanic herbalism (Box 1) with conventional medicine.

---

[1] Scherrer MM, Zerbe S, Petelka J, Säumel I. Understanding old herbal secrets: The renaissance of traditional medicinal plants beyond the twenty classic species? Front Pharmacol. 2023 Mar 24;14:1141044. https://doi.org/10.3389/fphar.2023.1141044. PMID: 37033626; PMCID: PMC10079881.

E. Ernst, *The Leopard Lily Project*, https://doi.org/10.1007/978-3-032-24771-1_6

An apt example of the widespread use of herbal medicine is the treatment of Adolf Hitler. In 1936, Hitler was suffering from chronic eczema, recurrent abdominal spasms, flatulence, exhaustion, and periods of depression. His photographer recommended Dr. Theodor Morell,[2] a Berlin physician for the high society, and arranged a personal meeting between the two. Morell then visited Hitler in Berchtesgaden. Morell's treatment appeared to be successful, and Hitler invited him to become his full-time personal physician and even arranged an honorary professorship for Morell. The treatments that Morell gave to Hitler were often herbal or natural and included:

- Franzbrandwein,
- Camomile tea,
- probiotics,
- various belladonna extracts,
- nux vomica,
- a preparation of bile extract, angelica, aloes, papaverine, caffeine, and pancreatin,
- a compound of enzymes, amino acids, and vitamins B1, B2, and C plus extracts of cardiac muscle, suprarenal, liver, and pancreas,
- a mixture of strophantin, glucose, vitamin B, and nicotinic acid,
- homeopathic Strophanthus gratus,
- preparation made from animal glands, livers, and other organs,
- vitamins A, D, and B12,
- an enzyme extract from Aspergillus oryzae.

In addition, Morell administered a range of powerful conventional drugs, and it is not unreasonable to assume that several of Hitler's health problems were caused by adverse effects of and/or interactions between multiple medications.[3]

One of the Nazi's main projects in relation to herbal medicine was the large-scale herb plantation at the Dachau concentration camp euphemistically called the "herb garden."

In 1939, Oswald Pohl (Chap. 13), the head of the "SS Main Economic and Administrative Office," founded the "German Research Institute for Nutrition and Food Provision Ltd." It had the following tasks:

---

[2] Theodor Morell: Die dubiosen Methoden von Hitlers Leibarzt—Opiate, Hormone, Aderlass—DER SPIEGEL.

[3] Adolf Hitler's Personal Charlatan | Skeptical Inquirer.

- systematic research and cultivation of medicinal herbs native to Germany in the interest of the national economy,
- supplying German and foreign markets with German drugs,
- production of new drugs and new syntheses based on scientific research,
- maintenance of laboratories,
- acquisition of plots,
- organization of all commercial and agricultural transactions arising in connection with the enterprise, e.g., poultry and animal farms, etc.

The plantation at the Dachau concentration camp was the centerpiece of this complex project. It had been initiated in 1938 under the control of the SS and was located outside the Dachau main camp. It formed part of the economic ventures under the "German Research Institute for Nutrition and Food Provisions Ltd." The hard labor was carried out almost exclusively by camp prisoners. Ernst-Günther Schenck (Box 2) was tasked to oversee the operation from the scientific and medical perspective. Among other things, vitamin supplements for the Waffen-SS were manufactured from these plants.[4] The aim was to render Nazi Germany self-sufficient in terms of its requirements for pharmaceuticals.

The plantation spanned 200 hectares and included greenhouses, a research institute (Box 3), a shop, and extensive arable land. Prisoners often worked in inhumane conditions, with inadequate clothing and food, leading to high mortality rates; around 800 deaths were recorded at the plantation. Some of the produce was sold to the public in a dedicated shop. Responsible for selling the produce and testing was the SS-owned company "Deutsche Versuch sanstalt für Ernährung und Verpflegung GmbH." The plantation included numerous further structures, including a maintenance building, a teaching and research institute, an equipment shed, a bee house, greenhouses, as well as large sections of productive land. Today, only the former administrative and institute buildings as well as some remnants of three greenhouses have survived.[5]

In 1941, Himmler was informed that the extracts of the plant, *Dieffenbachia seguine*, commonly called Leopard Lily, might be effective for inducing infertility. He understood that Leopard Lily might be a practical solution to the problem that he had long wanted to tackle: the mass sterilization of racially undesirable prisoners and people from the occupied territories in the

---

[4] Dachau herb gardenHistory of Sorts.

[5] SS experimental agricultural facility "herb garden"/ "plantation"—KZ Gedenkstätte Dachau.

East.[6] This idea started the Leopard Lily project which is the subject of the next chapters.

Box 1: Traditional Germanic Medicine

- Deeply intertwined with the natural environment and pre-Christian spiritual beliefs. Plants were not just seen as medicinal but also as having magical or sacred properties.
- Relied on a variety of local flora, e.g., yarrow, known for its wound-healing properties; nettle, used for a wide range of ailments from joint pain to internal issues.
- Knowledge was passed down orally, with specific individuals, often women known as "wise women" or healers, acting as the custodians of wisdom.
- With the arrival of Christianity, a significant part of the tradition was preserved and formalized in monasteries.
- The traditions were heavily influenced by prominent figures, e.g., Hildegard of Bingen (1098–1179).
- The legacy of traditional Germanic herbalism laid the foundation for modern phytotherapy in Germany.

Box 2: Ernst-Günther Schenck (1904–1998)

- Member of the SA and the NSDAP as well as numerous further NS organizations.
- In 1940, he became the Inspector for Nutrition of the Waffen-SS.
- He conducted dietary experiments on camp prisoners that led to several fatalities.
- In the last days of Hitler's life, Schenk met the Fuehrer twice in person.
- After the war, he was sentenced by the Soviets to a 10-year prison sentence.[7]
- He was released in 1953 and returned to West Germany.
- As his medical license had been withdrawn, he then worked, together with other former Nazi doctors, for the pharmaceutical company Grünenthal.
- Schenk's book "Patient Hitler"[8] was published in 2000.

Box 3: The "Research Institute" at the Plantation

---

6 Kenny MG. A darker shade of green: medical botany, homeopathy, and cultural politics in interwar Germany. Soc Hist Med. 2002 Dec;15(3):481–504. https://doi.org/10.1093/shm/15.3.481. PMID: 1265909.

7 Ernst-Günther SchenckWikipedia.

8 Patient Hitlerbk1311: Ernst Günther Schenck: Amazon.co.uk: Books Schenck E G: Patient Hitler. Weltbild Verlag (200).

- A primary goal was to optimize the cultivation of native German herbs to reduce reliance on imports.
- Large fields of gladioli were cultivated to extract vitamin C from the flowers for use by the armed forces.
- Investigations were carried out into anthroposophical biodynamic agriculture principles.
- Some of the products from the plantation were submitted to experimental testing on camp inmates. For example, a cream developed for frostbite protection by Weleda was supplied to Sigmund Rascher (Chap. 5, Box 3) for use in experiments on prisoners.

7

Leopard Lily

Dieffenbachia seguine, commonly known as Leopard Lily, is a highly toxic plant native to parts of South America and India. Contact with the plant's sap can cause severe irritation, burning, and swelling. This action is due to the oxalate crystals in the sap. Ingesting even a small amount of the plant can lead to a loss of speech for several days (hence also common name of the plant "dumbcane") and, in rare cases, can even be fatal.

Leopard Lily or Dumbcane or Poison Arum (Schweigrohr[1] in German) are the common names for *Dieffenbachia seguine*.[2] This popular ornamental plant is native to parts of South America as well as West- and East-India (Box 1). It is the most toxic genus of the Araceae family.[3] The genus is named after Joseph Dieffenbach,[4] a nineteenth-century botanist and gardener at the Botanical Gardens in Vienna. Confusion arises because other plants are also sometimes referred to as "Leopard Lily," e.g., *Lilium pardalinum* (Panther Lily), *Iris domestica* (Blackberry Lily), *Ledebouria socialis* (Silver Squill or Violet Squill).

---

[1] Dieffenbachie—Wikipedia.

[2] Dieffenbachia—Wikipedia.

[3] Araceae—Wikipedia.

[4] Joseph Dieffenbach—Wikipedia.

© The Author(s), under exclusive license to Springer Nature Switzerland AG 2026

E. Ernst, *The Leopard Lily Project*, https://doi.org/10.1007/978-3-032-24771-1_7

Accidental poisonings with Leopard Lily, particularly those of children, are reported with some regularity.[5] Ingestion of even a small portion of stem causes a burning sensation as well as severe irritation of the mouth, throat, and vocal cords. In severe cases, the swelling of the tongue and mouth can cause choking. When stems are bitten, the resulting stomatitis may be severe enough to bring about aphonia and render victims speechless for several days (hence the common names Dumbcane and Schweigrohr in German).[6]

> If one Cut this cane with a Knife, and put the tip of the Tongue to it, it makes a very painful sensation, and occasions such a very great irritation on the salivary ducts, that they presently swell, so that the person cannot speak, and do nothing for some time but void spittle in a great degree, or salivate, which in some time goes off, in this doing in a greater degree, what European Arum does in a lesser, and from this its quality, and being jointed this Arum is called Dumbcane.[7]

Severe corrosive lesions may occur in the mouth, larynx, esophagus, and stomach which may be complicated by respiratory failure, bradycardia, muscle twitching, cramps, vomiting, diarrhea, and hypertension. Deaths due to the toxicity of Leopard Lily have been reported.[8] On occasion, severely affected victims can be saved only through the insertion of a tube into the trachea.

When the sap of Leopard Lily enters the eye, it causes intense pain, watering, and blepharospasms. After about one hour, the eyelids become swollen.[5] The condition is caused by oxalate crystals (see below) injuring the cornea, the clear, protective outer layer of the eye. The condition usually resolves after a few days.[9]

There is no specific antidote against poisoning with Leopard Lily. The recommended treatments for acute oral intoxications include respiratory

---

[5] Niţescu GV, Grama A, Turcu T, Strătulă A, Dragomirescu A, Pană ES, Baciu A, Baconi DL, Crăciun MD, Ulmeanu CE. Epidemiology and Clinical Characteristics of Acute Plant Exposure in Patients Aged between 0 and 18 Years-A Six-Year Retrospective Study. Children (Basel). 2024 Feb 21;11(3):271. https://doi.org/10.3390/children11030271. PMID: 38,539,306; PMCID: PMC10969538.

[6] Arditti j, Rodriguez E: DIEFFENBACH/A: USES, ABUSES AND TOXIC CONSTITUENTS: A REVIEW. Journal of Ethnopharmacology, 5 (1982) 293–302.

[7] Sloane, H., A Voyage to the Islands of Madera, Barbados, Nieves, S. Christophers, and Jamaica, printed by B. M. for the Author, London, 1707.

[8] Dieffenbachia—an overview | ScienceDirect Topics.

[9] Chong SH, Ch'ng TW, Mustapha M. Dieffenbachia-Induced Transient Crystalline Keratopathy: A Case Report and Review of Previously Reported Cases. Cureus. 2022 Jan 12;14(1):e21146. https://doi.org/10.7759/cureus.21146. PMID: 35,165,597; PMCID: PMC8831483.

support, gastric lavage, water or milk by mouth, and antihistamines.[10] Eye injuries are best treated by rinsing with physiologic saline solution and sterile covering of the affected eye.[11]

Toxic Constituents and Modes of Action.

It is not entirely clear precisely which constituents of the Leopard Lily render this plant toxic. A generally held assumption is that the needle-like oxalate crystals contained in the sap cause mechanical micro-damage to tissues that encounter it. This tissue damage is thought to facilitate the entry and subsequent poisoning caused by oxalates, enzymes and other plant constituents.[12] The result is a release of histamine and kinin in the tissue which, in turn, causes the edema. This chain of events might explain the localized effects; however, it fails to account for the plant's systemic toxicity. It might involve also other constituents of Leopard Lily, including:

- saponins,
- phenolic compounds,
- proteins,
- enzymes,
- pigments.

Traditional Medicinal Uses.

Preparations containing Leopard Lily have been employed traditionally and are still recommended in some textbooks of herbal medicine for the treatment of a bizarrely wide range of conditions,[5,13] e.g.:

- angina,
- cancer,
- dropsy,
- dysmenorrhea,

---

[10] Mintzker Y, Bentur Y. [DIEFFENBACHIA POISONING]. Harefuah. 2018 Oct;157(10):631–633. Hebrew. PMID: 30,343,538.

[11] Bagatur Vurgun E, Arslan SC, Turhan SA, Toker AE. To report a case of crystalline keratopathy induced by Dieffenbachia plant sap and literature review. Am J Ophthalmol Case Rep. 2022 Jan 31;25:101,383. https://doi.org/10.1016/j.ajoc.2022.101383. PMID: 35,198,811; PMCID: PMC8850207.

[12] 1,030,474.

[13] Hagers Handbuch, Springer Verlag, Heidelberg, 1992.

- frigidity,
- gout,
- impotence,
- edema,
- varicose veins,
- warts.

According to one report, "The root is of more force than the fruit or leaves; besides, the first qualities, being of very small parts, and opening obstructions, fomentations are made of them against inflammation and obstructions of hypochindres and reins; and the oil is good against those evils, and supplies that of capers, and lilies. The roots are sliced and boiled in wine, made into baths, and used to the feet, it is of great use against old and late gouts."[5]

Due to the well-documented toxicity, most herbalists of today would think twice about prescribing remedies that contain Leopard Lily. Yet, in homeopathy, Leopard Lily is still used regularly (Box 2); several commercial products are currently available, for instance, for the treatment of intense itching or motion sickness.,[14,15]

Whether used as a herbal or homeopathic remedy, none of the therapeutic claims made for these remedies is backed up by sound evidence. In other words, the use Leopard Lily is not warranted as a treatment of any condition.

Sterility.

Numerous herbal medicines have been shown to affect fertility; some are known for their ability to increase fertility,[16] while others may cause infertility.[17] Leopard Lily belongs to the latter group. It had long been noted that the male inhabitants of Caribbean Islands followed the tradition of chewing Leopard Lily as a means of contraception. According to some reports, this habit brought about sterility lasting for 24–48 h.[18] Whether this traditional custom had been known to Gerhard Madaus when he embarked on his research into the effects of Leopard Lily on sterility, the subject of the next chapter is uncertain.

---

[14] Caladium seguinum 30C.

[15] CALADIUM SEGUINUM- dieffenbachia seguine | myHealthbox.

[16] 10 Plants That May Help Boost Fertility Naturally—Pregged.com.

[17] What Herbs Cause Infertility? The Risks Revealed—GardenerBible.

[18] Walter, W. G. and Khanna, P. N., Chemistry of the aroids. I. Dieffenbachia seguine, amoena andpicta. Economic Botany, 26 (1972) 364–372.

Box 1: Description of the Plant, Leopard Lily

- Grows 3–6 feet tall indoors and taller in ideal outdoor conditions.
- Toxic when ingested.
- An irritant to the skin.
- Prefers bright, indirect light, well-draining soil, and moderate watering.
- For safe handling it is best to wear gloves when touching the plant.
- Keep the plant out of reach of children and pets.

Box 2: Leopard Lily as a Homeopathic Remedy[19]

- Homeopathic remedies are normally highly diluted; therefore, the plant's toxicity may be irrelevant.
- It is recommended to treat sexual disorders of psychological origin, especially in men.
- It is also used for intolerance to mosquito bites and those of other insects.
- In women, it is claimed to be an effective remedy for genital pruritus, affecting the vulvar region and vagina.
- It is furthermore said to cure addiction problems and, in particular, tobacco addiction.
- None of these claims is supported by reliable evidence which sadly does not stop homeopaths from repeating them.

---

[19] Caladium Seguinum Homeopathy

8

Gerhard Madaus (1890–1942)

Gerhard Madaus studied medicine and, along with his brothers Friedemund and Hans, founded the company Dr. Madaus & Co. specializing in plant-based remedies. Gerhard sought to scientifically evaluate herbal medicines. Among other subjects, he studied herb-induced sterilization in rats. While his company faced allegations of complicity in unethical Nazi experiments, these claims were later rejected in court, although controversy remained.

Gerhard Madaus was born in Nestau, Germany. His mother, Magdalene Madaus (1857–1925), was a non-medical herbalist, homeopath, and iridologist (Box 1).[1] She and her husband, Heinrich Friedrich Pieter Madaus, an Old Lutheran pastor, had seven children: Eva, Susi, Hanna, Gerhard, Agnes, Friedemund, and Hans. Health challenges led Magdalene to seek treatment from Pastor Emanuel Felke[2] who treated Magdalene's pelvic inflammation and Hans's polio with homeopathy. Magdalene was most impressed with this success and became Felke's student. In 1908, she opened her own clinic where she treated patients mainly with homeopathy. She manufactured her own remedies formulated from plants, minerals, and organic compounds which became known as "oligoplexes."[3] She published several influential textbooks, her contributions were internationally recognized, and she even earned an

---

[1] Magdalene Madaus – Wikipedia.

[2] Emanuel Felke - Wikipedia.

[3] Oligoplexe Madaus - Homöopathie - Forum für Naturheilkunde & Alternativmedizin - Yamedo.

© The Author(s), under exclusive license to Springer Nature Switzerland AG 2026

E. Ernst, *The Leopard Lily Project*, https://doi.org/10.1007/978-3-032-24771-1_8

honorary doctorate from the "American School of Naturopathy" in New York.

Gerhard Madaus studied medicine and graduated in 1919. In the same year, he and his two brothers—Friedemund, a banker, and Hans, a pharmacist—founded the pharmaceutical company Dr. Madaus & Co. The company initially produced "Biopastillen" (organic pastilles), plant-based pills aimed at providing natural health solutions.

The company proved to be highly successful and expanded rapidly. Due to the French occupation of the Rhineland in 1921, it moved its headquarters to Radeburg in Saxony. Soon after, it opened further branches in:

- Stuttgart,
- Radeburg,
- Berlin,
- Amsterdam,
- Warsaw.

At the time, herbalism (Chap. 6) was mostly in the hands of lay practitioners who had little interest in science. Gerhard Madaus was therefore unusual: He had studied medicine, wanted to scientifically validate his remedies and developed innovative methods, for example, techniques for extracting active compounds from plants without heat.[4]

In 1938, Madaus published his 11-volume "Lehrbuch der Biologischen Heilmittel" (Textbook of Biological Medicine, Box 2).[5] It aimed to legitimize herbal medicine scientifically and gained considerable attention, not least by Nazi officials who recognized Madaus as a leading expert in the field of scientific herbal medicine. Little is known about Madaus political convictions; he was not a member of the NSDAP but had joined the "Stahlhelm" (Box 3).[6]

In 1941, Madaus and his co-worker Friedrich Koch (Chap. 9) published an article in a German science journal that would have far-reaching consequences. Its title was "Tierexperimentelle Untersuchungen zur Frage der

---

[4] Timmermann C. Rationalizing 'folk medicine' in interwar Germany: faith, business, and science at "Dr. Madaus & Co.'. Soc Hist Med. 2001 Dec;14(3):459–82. https://doi.org/10.1093/shm/14.3.459. PMID: 11,811,189.

[5] Lehrbuch der biologischen Heilmittel - 11 Bände (Nachdruck des Originals von 1938): Gerhard Madaus: Amazon.de: Books Madaus G: Lehrbuch der biologischen Heilmittel. OLMS re-edition (2021).

[6] Der Stahlhelm, Bund der Frontsoldaten - Wikipedia.

medikamentoesen Sterilisierung" (Experimental animal studies on the question of drug-induced sterilization).[7] In their paper, the authors mention that natives have used Leopard Lily (they use the Latin name "*Caladium seguinum*") for rendering their enemies infertile. They also report that previous experiments on mice had yielded inconsistent results. This, they suggested, might have been due to the inconsistent quality of the raw product used.

Therefore, Madaus and Koch decided to conduct further experiments. The researchers used 12 laboratory rats with proven fertility and observed that they invariably were rendered infertile through the oral or subcutaneous administration of 0.5 ml freshly pressed juice of Leopard Lily leaves per day. The time required to achieve sterility was 30–50 days for female and 40–90 days for male rats. These effects were accompanied by a general atrophy of the animals' sexual organs. The authors also mentioned in passing and without providing data that they had similar results after administering Leopard Lily to both rabbits and dogs. They concluded their article by stating that it would be important to investigate whether Leopard Lily caused similar effects in humans. At no stage, however, did they advocate using the plant for mass sterilization of humans, and there is no suggestion that they viewed their findings in relation to the Nazi's sterilization projects.

Yet, some sources claim that Madaus became indirectly involved in unethical and cruel experiments conducted with the plant Echinacea on Buchenwald concentration camp prisoners. The aim allegedly was to test its efficacy against burns from phosphorus bombs. From November 1943 to January 1944, prisoners were subjected to burns using phosphorus from incendiary bombs to test various treatments, causing severe pain and injury.

In 1943, a doctor in the Buchenwald concentration camp tested the effectiveness of the drug Echinacin (echinacea) on five inmates who had been burned with a phosphorus-rubber mixture—inspired by successful experiments with echinacea extracts on animals in the laboratories of the company Dr. Madaus & Co. in Radebeul near Dresden.[8]

Provided that the mention of the year 1943 is correct, Gerhard Madaus cannot have been directly involved because he had died in 1942 in Dresden. Whether Koch (Chap. 9), who continued the research of Madaus after his death, was involved seems unclear.

---

[7] Madaus G, Koch F E: Tierexperimentelle Untersuchungen zur Frage der medikamentoesen Sterilisierung. Zeitschrift fuer die gesamte experimentelle Medizin einschliesslich Chirurgie, 109, 68–87, 1941.

[8] StadtRevue Köln Magazin | Archiv.

The events of the war and the collapse of the Third Reich severely disrupted the Madaus company's operations. Several of the firm's facilities were destroyed, and the Radebeul site was eventually expropriated and dismantled. The company's active involvement in any unethical Nazi activities remained unproven. In 1949–1950, the company, meanwhile relocated in West Germany, successfully sued the newspaper "Volksstimme" for alleging the firm's complicity in Nazi experiments; the paper was henceforth prohibited from repeating such claims.

After Friedemund and Hans Madaus had re-established the company in West Germany, it expanded rapidly by adding new production sites and became one of Europe's leading manufacturers of herbal preparations. In 1992, Andreas Madaus, a grandson of Gerhard Madaus, joined the company. The German healthcare reforms of the early 1990s led to a significant drop in revenue, and the workforce had thus to be reduced from 800 in 1994 to about 400 by 1999. In 2001, Madaus was sold and is today part of a large international conglomerate.

Box 1: Magdalene Madaus

- Magdalene Johanne Marie Madaus, née Heyer, was born in 1857 in Magdeburg, Germany.
- She often suffered health issues, including several miscarriages, and her children frequently fell ill. Her son Hans suffered from polio.
- She consulted Emanuel Felke[9] which was followed by her recovery and even her son Hans learned to walk again.
- Subsequently, Magdalena became a naturopath and a homeopath.
- She developed her own system of "complex homoeopathy" and published several books about it.
- Her work had an important influence on her son, Gerhard Madaus.
- Magdalene Madaus died in 1925 in Dresden, Germany.

Box 2: Gerhard Madaus' Textbook

- The book is a comprehensive encyclopedia of medicinal plants, documenting a vast number of species and their historical and traditional uses.
- It advocates for a holistic and biological approach to health care.
- It is an attempt to bridge the gap between traditional folk medicine and modern science.
- Each volume contains detailed monographs on individual plants, including their botanical description, chemical constituents, known actions on the body, and therapeutic applications.

---

[9] Bear Foot Doctor's health blog - BEAR FooT DOCTOR ® In Step With Nature.

- The book became a foundational text for the natural healing movement in Germany and influenced the practice of biological medicine for decades.
- It remains a significant historical reference in the field of herbal medicine and is still cited today.

Box 3: "Der Stahlhelm, Bund Der Frontsoldaten" (the Steel Helmet, League of Front-Line Soldiers)

- A major German veterans' organization active during the Weimar Republic and early Third Reich founded 1918 by World War I veterans.
- Its ideology was ultranationalist, anti-democratic, and monarchist, rejecting both the Treaty of Versailles and the Weimar Republic.
- It evolved into a massive, politically active paramilitary group. By the early 1930s, it had a membership of around 500,000 men.
- It aimed to destabilize the parliamentary republic and replace it with an authoritarian or military-backed state.
- The Stahlhelm leadership mistrusted Hitler.
- The Stahlhelm and the NSDAP were initially rivals.
- Despite the rivalry, the Stahlhelm briefly joined the Nazis in 1931 to form the Harzburg Front, a temporary political alliance against the Weimar government.
- After Hitler came to power in January 1933, the Stahlhelm was renamed the Nationalsozialistischer Deutscher Frontkämpferbund (National Socialist German Front-Fighters' League).
- In 1935 it was absorbed by SA.

9

Friedrich Koch (1901–1985)

Friedrich Ernst Koch was a co-worker of Gerhard Madaus conducting research on herbal medicine, including herb-induced sterilization of animals using Leopard Lily. Their work caught the attention of Himmler, who wanted to use this remedy for the mass sterilization of humans. In his post-war testimony, Koch claimed that he intentionally slowed down this research to thwart Himmler's plans. Koch was thus permitted to continue as a researcher for the Madaus company.

Friedrich Ernst Koch was born in Fulltau, Germany.[1] He studied medicine and specialized as a pharmacologist. He then became the director of the Biological Institute at Dr. Madaus & Co. (Chap. 8) conducting scientific research aimed at validating the efficacy of medicinal plants.[2]

Koch collaborated closely with his boss, Gerhard Madaus, the institute's founder, on research into a range of herbal remedies. Their work focused on animal experiments, including the ones on the sterilization of rats through the administration of extracts of Leopard Lily.

In October 1941, Dr. Adolf Pokorny (Chap. 10) came across a scientific paper by Madaus and Koch. He then wrote to Himmler (Chap. 5) about

---

[1] Harvard Law School Nuremberg Trials Project, metadata associated with Koch's affidavit in the Doctors' Trial (HLSL Seq. No. 784).

[2] Klee E: Das Personenlexikon zum Dritten Reich. Fischer, Frankfurt, 2015.

E. Ernst, *The Leopard Lily Project*, https://doi.org/10.1007/978-3-032-24771-1_9

this research and its potential. Himmler initially ignored the letter, presumably because Pokorny was not a member of the Nazi elite. He thus took some time to grasp the relevance of Pokorny's letter. It was only on March 10, 1942, that Oswald Pohl (Chap. 13) received an order from Himmler to contact Madaus. Pohl was instructed to persuade Madaus, who unbeknown to Himmler had recently died, to conduct sterilization experiments with Leopard Lily on humans. The plan was to initiate this research in cooperation with the SS Reich Physician Ernst-Robert Grawitz (Chap. 14). The experiments were to be carried out on criminals who were already subject to court-ordered sterilizations.

On June 3, 1942, Pohl reported to Himmler about the progress on the Leopard Lily project indicating that he had been in contact with Koch who considered the sterilization experiments "extremely important." The SS then instructed the Madaus Company to conduct further research as a matter of urgency.

Yet, progress turned out to be slow, as Pohl reported: "caladium seguinum grows only in North America* and during the war, could not be exported in adequate quantities." Pohl further explained that "attempts to grow the plant from seed cultivated in hot houses made by Koch had been successful. However, growing the plant and developing the drug was not fast enough and the yield not sufficient to permit large scale experimentation." A reply soon came from Rudolf Brandt (Box 1),[3] Himmler's the Personal Administrative Officer, assuring Pohl that "a large hot-house will be placed at Koch's disposal as soon as possible" and that Himmler considered the experiments: "extremely important!"[4]

To check on Koch's experiments, the SS sent Karl Tauböck,[5] a chemist from the biological laboratory of IG Farben to the Biological Institute of the Madaus Company in Radebeul. Tauböck subsequently reported that, although sterilization with Leopard Lily was not ideal, it was possible.[3] Koch was personally involved both in the discussions which were supervised by SS-Obersturmbannführer Dr. Lolling (Box 2), as well as in the research itself.[1]

After the war, Koch appeared as a defense witness in the Nuremberg Doctors' Tribunal (Chap. 10, Box 1) and provided an affidavit. Koch's testimony was part of the defense for the defendant Adolf Pokorny. Koch insisted

---

[3] Rudolf Brandt—Wikipedia.
[4] 'Three Million Bolsheviks Could Be Sterilised' | Project | Nuremberg. Casus pacis.
[5] Nuremberg—Karl W. F. Tauboeck.

that nobody at the Madaus Company had ever handed over Leopard Lily related material to the SS or to other Nazi officials. He further stated that he intentionally slowed down the progress of these experiments. His hope had been that the interest in mass sterilization with Leopard Lily would fade. For his delay strategy, he gave the following motivation: "The reason for this is that we suspected that the SS or Pohl might have intentions with which we did not agree."[6]

While Koch was directly involved in discussions with several SS officials about Leopard Lily experiments and most likely received financial support from the SS for carrying out the research, there is no evidence to assume that he personally conducted experiments with Leopard Lily on humans. His role appears to have been entirely research oriented. Yet, one could nevertheless argue that his cooperation with the SS implicates him in the broader unethical framework of unethical research.

Koch was never accused of any wrong doings committed during the Third Reich. He was able to resume his research for the Madaus company after the war.[7] He died in 1985 in Geesthacht.[2]

*This was an error by Pohl: Leopard Lily is native to South America.

Box 1: Rudolf Hermann Brandt (1909–1948)[8]

- A lawyer and a stenographer by profession. His skills were noticed by Himmler, who transferred him to his staff.
- He became Himmler's Personal Administrative Officer, handling all of his correspondence except for matters relating to the Waffen-SS or the Police.
- Brandt was also a member of the Ahnenerbe, of which Himmler was president.
- Brandt was responsible for the administration and coordination of several experiments at concentration camps, including the Jewish skull collection.
- After the war, Brandt was imprisoned and became a defendant in the Nuremberg Doctors' Trial.
- He was tried for his role in medical experiments.
- He was found guilty of war crimes, crimes against humanity, and being a member of a criminal organization, i.e., the SS.
- Brandt was sentenced to death and executed by hanging on June 2, 1948, his 39th birthday.

---

[6] Mitscherlich A, Mielke F: Medizin ohne Menschlichkeit. Fischer, Frankfurt, 1993.

[7] Erbring H, Koch E, Lorenz D, Madaus R, Orzechowski G, Patt P. Therapie mit nativen Convallaria-Glykosiden (NCG); chemische, pharmakologische und klinische Betrachtungen [Therapy with native Convallaria glycosides; chemical, pharmacological & clinical observations]. Medizinische. 1958 Aug 30;35:1326–31. German. PMID: 13577370.

[8] Rudolf Brandt—Wikipedia.

Box 2: Dr. Enno Lolling (1888–1945)[9]

- Joined the SA in 1923, the NSDAP in 1931, and the SS in 1933.
- September 1936, SS squadron doctor with the SS-Verfügungstruppe at the SS military academy in Bad Tölz.
- November 1936, he became a doctor at the SS military hospital in Dachau.
- 1941, camp doctor at the Dachau concentration camp.
- Same year, chief camp physician at Sachsenhausen concentration camp.
- Same year, chief physician at the Concentration Camps Inspectorate.
- 1942, in charge of Amt D III of the SS-Wirtschafts-Verwaltungshauptamt for Medical Services and Camp Hygiene.
- After the war, Lolling committed suicide.

---

[9] Enno Lolling—Wikipedia.

10

Adolf Pokorny (1895–??)

Adolf Pokorny, an Austrian dermatologist, wrote to Himmler in 1941, suggesting that Leopard Lily could be used for the mass sterilization of "three million Bolsheviks." Himmler decided to pursue this idea and thus initiated the Leopard Lily project. Pokorny himself had no further involvement in this initiative. After the war, he was indicted at the Nuremberg Doctors' Trial but was acquitted because there was no evidence that Pokorny's suggestions of human experiments were ever carried out.

Adolf Pokorny was born in Vienna as the son of an Austrian military officer, Josef Pokorny and his wife Kristina.[1] He had one sister, Olga, and the family spent their childhood in Bohemia, Galicia, and Bosnia. Adolf Pokorny served in WW I and then studied medicine in Prague. He received his license to practice medicine in 1922, then worked in various hospitals, and, from 1924, practiced as a doctor in private practice in Komotau, Czech Republic.[2] Sources differ as to when in his life he had adopted which nationality; in most documents, however, he is identified as an Austrian.[3]

In 1923, Adolf Pokorny married the Jew Lilly Weil.[4] The couple had a son and a daughter. The marriage ended in divorce in 1935. After the German

---

[1] Adolf Pokorny—Biographical Summaries of Notable People—MyHeritage.

[2] Adolf Pokorny | AustriaWiki im Austria-Forum.

[3] Nuremberg—Adolf Rudolf Pokorny.

[4] Lilly Pokorná—Wikipedia.

E. Ernst, *The Leopard Lily Project*, https://doi.org/10.1007/978-3-032-24771-1_10

annexation of the Sudetenland in October 1938, the Nazis tried to confiscate half of his house, which had originally belonged to his wife. In 1939 Pokorny applied to join the NSDAP but was refused due to his previous marriage with a Jew. During WW II, Pokorny worked as a medical officer.[5]

Lilly Weil was a Czech-Brazilian radiologist who, after their divorce, had moved to Prague (Box 1). She managed to send her two children to England via a "Kindertransport." In March 1942, she was deported to the Theresienstadt ghetto (Box 2) where she established and directed the first radiology unit. She survived the war and thereafter emigrated to São Paulo, Brazil, where she died in 1974.[6]

In October 1941, Pokorny wrote a letter to Himmler (Chap. 5): "Driven by the idea that the enemy must not only be defeated, but destroyed," he suggested to carry out sterilization experiments with Leopard Lily. Pokorny's letter referred to a publication by Madaus (Chap. 8), which seemed to demonstrate that the sap of this plant caused permanent sterility in experimental animals. In Pokorny's own words[7]:

Dr Madaus published the result of his research on medical sterilization. Reading these articles, the immense importance of this drug in the present fight of our people occurred to me. If, based on this research, it was possible to produce a drug, which after a relatively short time, affects imperceptible sterilization on human beings, then we would have a new powerful weapon at our disposal.

The thought alone that the 3 million Bolsheviks—at the time German prisoners—could be sterilized so that they could be used as laborers but be preserved from reproduction opens the most far-reaching perspectives.

Madaus found that the sap of the caladium seguinum, when taken by mouth or given as an injection to male and female animals produces, after a slight delay, permanent sterility. The illustrations accompanying the scientific article are convincing.

If the ideas meet with your approval, the following course should be taken:

(1) Dr Madaus must not publish any more such articles (the enemy listens!),
(2) cultivation of the plant (easily done in green houses!),
(3) immediate research on human beings (criminals!) to determine the dose and length of the treatment,

---

[5] 1947-08-19, #23: Doctors' Trial Verdict—Dr. Adolf Pokorny (substack.com).

[6] Dr. med. Lilly Pokorná (Weil) (1894–1974)—Genealogy.

[7] 'Three Million Bolsheviks Could Be Sterilised' | Project | Nuremberg. Casus pacis.

(4) quick research of the formula of the effective chemical substance to produce it synthetically, if possible.

> If it was possible to produce a drug that would generate undetected sterilization in humans in a relatively short time, we would have an effective weapon at our disposal. The mere thought that the three million Bolsheviks currently in German captivity could be sterilized so that they would be available as workers but excluded from reproduction opens up far-reaching prospects.

At the time, Pokorny was a reserve medical officer, by no means part of the Third Reich's medical elite and completely unknown to Himmler. He lived in a provincial Sudeten town and, for all intents and purposes, was irrelevant to the Nazis. This perhaps explains why Himmler initially ignored Pokorny's letter. When he finally realized its relevance, he considered this information to be valuable and instructed Oswald Pohl (Chap. 13) and Ernst-Robert Grawitz (Chap. 14) to follow up Pokorny's suggestions.

Even though Pokorny's letter was then given the utmost importance and his suggestions were pursued with extreme urgency, Pokorny himself seemed to have had no further involvement in any of the events that followed. In fact, he did not even receive a commendation or other acknowledgment from Himmler or other SS officials for his initiative.

After the war, Pokorny's letter was found in Himmler's papers. Pokorny was then promptly arrested and indicted in the Nuremberg Doctors' Trial of 1946 (Chap. 15). Pokorny defended himself by arguing that he had been aware all along of the ineffectiveness of Leopard Lily. He had written the letter merely to dissuade Himmler from using cruel methods of sterilization (Chap. 2). In other words, he tried to present his letter as an attempt not to bring about but to prevent harm to Nazi-victims.

The indictment stated that: Pokorny "claimed in his defence that he only wanted to prevent the execution of Himmler's intentions of mass sterilization and extermination of the population of the eastern territories and the Jews… He considered the idea of the sterilization by Leopard Lily unscientific and impossible of execution; that the conclusions of Madaus' articles did not quite correspond to the facts."[9]

The trail also raised the question as to how Pokorny could have possibly known about Himmler's plans of mass sterilization, given that they were top secret. Allegedly, a certain SS officer by the name of Voigt had consulted Pokorny in his practice in mid-1941. During the consultation, this patient happened to pick up a medical journal lying on the table of the waiting room. In it, he noticed the article by Madaus. Voigt then told Pokorny that he should bring this paper to Himmler's attention because he knew that the SS was currently looking for a way of mass sterilization by radiation. Pokorny then decided to write to Himmler in an attempt to prevent this atrocity.[8]

The court did not accept his argument, but nevertheless acquitted Pokorny: "We are not impressed by the defence put forward by the defendant, and it is difficult to believe that he was guided by the noble motives he claims when he wrote the letter. Rather, we are inclined to believe that Pokorny wrote the letter for entirely different and more personal reasons. [...] In Pokorny's case, the prosecution has not succeeded in proving his guilt. Outrageous and base as the suggestions in this letter are, there is not the slightest evidence that any steps were ever taken to apply them by experimentation on human beings. We therefore declare that the defendant must be acquitted, not because of, but in spite of the defence he has put forward."[9]

After his release, Pokorny would have been allowed to continue practicing medicine without restrictions. There is, however, no record to show that he did. It is therefore most likely that he did not return to a prominent medical career. Rather, he appears to have lived quietly and to have faded into obscurity.

Box 1: Lilly Pokorna/Weil (1894–1974)[10]

- Full name: Lilly Pokorná (née Weil).
- Lilly was born in Prague.
- She was married to Adolf Pokorny in 1922.
- The couple had a daughter, Lotte Christina Katz de Castro, and a son, Thomas Pokorny.[11]
- Both children survived the war in the UK.
- The couple divorced in 1935.
- Adolf Pokorny got custody of Thomas and Lilly from Lotte.
- Lilly was a physician and a radiologist.
- After her divorce, she worked in a dispensary for the "Society for the Fight against Veneral Disease."
- During her internment in Theresienstadt, she led the camp's first radiology station.
- After the war, Lilly emigrated with her children to Brazil.
- Lilly died in 1974 in São Paulo, Brazil.

---

[8] 'Three Million Bolsheviks Could Be Sterilised' | Project | Nuremberg. Casus pacis.

[9] Medizin ohne Menschlichkeit: Dokumente des Nürnberger Ärzteprozesses: Mitscherlich, Alexander, Mielke, Fred: Amazon.co.uk: Books Mitscherlich A, Mielke F: Medizin ohne Menschlichkeit. Fischer Verlag (1989).

[10] Dr. med. Lilly Pokorná (Weil) (1894–1974)—Genealogy.

[11] Thomas Pokorny (1929–2009)—Genealogy.

Box 2: The Ghetto of Theresienstadt (Terezín)[12]

- It operated from November 1941 to May 1945 in Terezín, in the German-occupied Protectorate of Bohemia and Moravia (now the Czech Republic).
- It functioned simultaneously as a transit camp for Jews being sent to extermination camps and as a labor camp for "privileged" Jews, e.g., decorated WWI veterans, artists, scholars, etc., from Germany, Austria, and Western Europe.
- The Nazis propaganda falsely portrayed it as a "spa town" or "model ghetto" where elderly Jews could "retire."
- Despite this façade, conditions were horrific due to extreme overcrowding, malnutrition, and disease, leading to a high death toll.
- Of the approximately 140,000 Jews transported to Theresienstadt, about 33,000 died within the ghetto, and nearly 90,000 were deported to killing centers like Auschwitz-Birkenau.
- The ghetto was liberated by Soviet troops on May 8, 1945.

---

[12] Theresienstadt Ghetto—Wikipedia.

11

Richard Rössler (1897–1945)

Richard Rössler was an Austrian pharmacologist who made significant contributions to medicine. Following the Anschluss, he took over his Jewish mentor's position and helped him emigrate. Rössler was an anti-Semite and a Nazi who collaborated with others on unethical research, including the Leopard Lily project. Rössler died at the age of 47 from a fatal head injury the cause of which remains unclear.

Richard Rössler was born in 1897 in Austria. He studied medicine at the University of Innsbruck and graduated in 1922. He then worked at the Institute for General and Experimental Pathology in Innsbruck (1921 1923) and the Pharmacological Institute in Graz (1923–1924). From 1924 to 1936, he served as an associate doctor under his Jewish mentor, Prof Ernst Peter Pick,[1] at the Pharmacological Institute of the University in Vienna.

Following the Anschluss in 1938, Pick was dismissed and Rössler was appointed as his successor. On March 12, 1938, the day of Hitler's arrival in Austria, Rössler is reported to have said: "I am afraid to go before my revered teacher Professor Pick and tell him that he has been dismissed and that I am his successor. How on earth am I supposed to do that? How to tell him that after a lifetime of leading and caring for every member of staff, he is no longer allowed to enter his own institute and other university buildings?"

---

[1] Ernst Peter Pick, Prof. Dr. | 650 plus.

E. Ernst, *The Leopard Lily Project,* https://doi.org/10.1007/978-3-032-24771-1_11

In the weeks that followed, Rössler took significant personal risks to help Pick leave Austria and secure his property.[2]

Rössler's research was mainly focused on the effects of drugs on the cardiovascular system and on bronchial resistance. This led to the development of isoprenaline, a breakthrough for the treatment of asthma. He also collaborated with the German firm Boehringer to test catecholamines, thus contributing to advancements in adrenoceptor classification. In 1941, Rössler was honored for his achievements by being elected a corresponding member of the Austrian Academy of Sciences in Vienna.

Rössler had been a member of the "German National People's Party" (Deutschnationale Volkspartei, Box 1)[3] and joined the NSDAP soon after the Anschluss. His official files of the time stated that he "stood ideologically on the ground of National Socialism during the illegal period" and "always displayed his anti-Semitic attitude."[1] From July 1944, Rössler collaborated with the SS physician Hermann Druckrey (Box 2), Director of the Pharmacological Institute of the Vienna Police Hospital VII.[4] The precise nature of their collaboration is not known.

Rössler also collaborated with Dr. Franz Fehringer (Chap. 12) over the Leopard Lily project. On October 14, 1942, another letter about the possibility of mass sterilization with Leopard Lily reached Himmler. It came from SS-Oberführer Karl Gerland (1905–1945).[5] He reported on the willingness of Dr Fehringer, head of the Regional Office for Racial Policy, to conduct studies on the effectiveness of Leopard Lily in Austria.

The response from Oswald Pohl (Chap. 13) explained that such studies were already underway. This information did, however, not stop Gerland. He nevertheless instructed Fehringer to go ahead and collaborate with Rössler to synthesize the active ingredients of Leopard Lily. In addition, Fehringer planned to cultivate Leopard Lily in greenhouses and to conduct experiments with Leopard Lily on the inmates of the Lackenbach "Gypsy" Camp (Box 3). Gerland requested the approval of his superiors to employ his staff together with Enno Lolling (Chap. 9, Box 2), head of the medical service at the concentration camps. Thus, a second project group started work on the possibility of employing Leopard Lily for mass sterilization of humans.[6,7]

---

[2] Richard Rössler (Mediziner)—Wikipedia.

[3] German National People's Party—Wikipedia.

[4] Hermann Druckrey—Wikipedia.

[5] Karl Gerland—Wikipedia.

[6] Klee E: Das Personenlexikon zum Dritten Reich, Fischer, Frankfurt, 2015.

[7] Mitscherlich A, Mielke F: Medizin ohne Menschlichkeit. Fischer Verlag (1989).

Little is known about the work that this second project group subsequently conducted. All attempts to synthetize the active ingredients seemed to have failed. Whether the experiments in the concentration camp ever took place seems doubtful.[8]

Richard Rössler died aged 47 on May 4, 1945, in Vienna, only weeks after the Red Army had captured the city. He had sustained a serious head injury but still managed to reach his institute where he died of his injury. To this day, it is unclear what had caused his injury. Some speculate that his death was caused by a suicide attempt prompted by his Nazi past.

Box 1: The German National People's Party (DNVP)

- Founded in 1918, the DNVP was a nationalist and monarchist political party during the Weimar Republic.
- It represented the interests of traditional German elites and rejected the Treaty of Versailles.
- After 1928, the party adopted a more radical nationalist stance and got closer to the NSDAP.
- It performed well in elections and became one of the largest parties in the Reichstag.
- In January 1933, the DNVP and the Nazis formed a short-lived coalition government. This partnership was key to Hitler's appointment as chancellor.
- The DNVP was dissolved in June 1933 under pressure from the Nazi regime, which by then had consolidated its power and no longer needed it.
- Many of its members subsequently joined the NSDAP.

Box 2: Hermann Druckrey (1904–1994)[9]

- Born in Greifswald, Germany, into a family of pharmacists.
- He completed a pharmaceutical apprenticeship before deciding to devote himself to medicine.
- Druckrey was an ardent Nazi; he joined the NSDAP and the SS in 1931.
- He became the leader of the "Nationalsozialistischer Deutscher Dozentenbund" (National Socialist Association of University Lecturers) from 1937 to 1942.
- He was awarded the Golden Party Badge, an award given to the oldest and most loyal members of the NSDAP.
- After the war, he was imprisoned from 1946 to 1947 in an allied internment camp.
- Druckrey achieved an outstanding international reputation and received numerous honors.

---

[8] 'Three Million Bolsheviks Could Be Sterilised' | Project | Nuremberg. Casus pacis.

[9] Hermann Druckrey—Wikipedia.

Box 3: The Lackenbach Gypsy Camp

- The camp was set up in November 1940.
- It was under the control of the criminal investigation department.
- The prisoners were used for forced labor.
- Roma were deported from the Lackenbach intermediate camp to concentration camps, other collection camps, or extermination camps.
- Part of the camp also served as a collection camp for Jewish Austrian prisoners.
- The number of prisoners fluctuated between 200 and 900, a third of whom were children.
- From the spring of 1941, the number rose to ~ 2000.
- The consequence was extreme overcrowding.
- When a typhus epidemic broke out in the winter of 1941/42, the prisoners were left to their fate with no medical care.
- The camp was liberated by the Red Army in April 1945.

12

Franz Fehringer (1903–??)

Franz Fehringer was a physician who served as the head of the Office for Racial Policy in Vienna. In this role, he was instrumental in the persecution of Jewish people and other groups deemed "undesirable," overseeing their identification, property confiscation, and deportation. He also worked as a T4 expert and was involved in the Leopard Lily project. His fate following the war is unknown.

Franz Fehringer was born in Fridau, Austria. He studied medicine in Vienna, joined the NSDAP in 1931, and became a member of the SA a year later.[1]

In September 1940, he was appointed head of the Office for Racial Policy (Box 1) in Vienna. Initially, the office was part of the NSDAP administration in the Vienna district and was responsible for propaganda, population policy, and the marginalization of entire population groups such as Jews, Roma, and Sinti, the disabled and people classified by the Nazi regime as so-called "asocials."[2] Fehringer thus oversaw measures that supported the identification, expropriation, and deportation of Jews. After the Anschluss, the Nuremberg Law extended to Austria and defined Jews not by religious but by racial criteria and stripped them of civil rights. Fehringer contributed to enforcing decrees requiring Jews to adopt additional names (e.g., "Israel" for men, "Sara" for women) and to register their property for confiscation and "Aryanization."

---

[1] Klee E: Das Personenlexikon zum Dritten Reich. Fischer Verlag Frankfurt. 2015.

[2] Rassenpolitisches Amt der NSDAP—Wien Geschichte Wiki.

E. Ernst, *The Leopard Lily Project*, https://doi.org/10.1007/978-3-032-24771-1_12

Fehringer worked in concert with other Nazi organizations, such as the "SS Race and Settlement Office" and the "Institute for the Study of Racial Hygiene," which conducted racial examinations of Jews, Roma, and Sinti. His role involved coordinating with Josef Bürckel, a key Nazi official in Vienna from 1938 to 1940, who aggressively promoted anti-semitic laws and property confiscation (Box 2).[3]

From September 1940 to January 1941, Fehringer also worked as an expert for the "Aktion T4."[4] This program, euphemistically called "euthanasia program," ran from 1939 to August 1941 and murdered approximately 70,000–80,000 people with physical or mental disabilities in Germany and Austria. It specifically targeted children and adults deemed "unworthy of life." T4 experts were typically physicians who reviewed the registration forms sent in from psychiatric institutions to decide who would be transferred to one of the killing centers like Hadamar, Hartheim, or Grafeneck. There victims were gassed and cremated. The program was stopped due to pressure from the public, but the killing continues under the heading of "wild euthanasia."[5]

In 1941, Fehringer began collaborating with Richard Roessler (Chap. 10), Adolf Pokorny (Chap. 9), and the Madaus company (Chap. 8) over the Leopard Lily project. On August 24, 1941, Karl Gerland had written to Himmler. His letter is referenced in Document NO. 039 from the Nuremberg Doctors Trial:

> ... Since one of the most urgent tasks of our National Socialist racial and population policy is to prevent the reproduction of the hereditarily unfit and racially inferior, the current head of the Gau Office for Racial Policy, Gau Headquarters Dr Fehringer, has dealt with questions of infertility and established that the previous options of castration and sterilisation alone did not produce the desired or intended success.[6]

The letter then referred to the Madaus experiments (Chap. 8) and suggested using Leopard Lily for that purpose. It is remarkable not least because it claimed things that are freely invented. Firstly, it claimed that Madaus and Koch conducted their experiments with Leopard Lily "in homeopathic doses," i.e., in highly dilute form. In truth, they used concentrated herbal tinctures. Secondly, the letter noted that the effects were demonstrated in "extensive research series on rats, rabbits and dogs." In truth, Madaus and Koch reported only a very limited experiment on rats and mentioned other tests merely in passing without providing any data at all.

---

[3] Josef Bürckel—Wikipedia.

[4] Klee E: Euthanasie im Dritten Reich. Fischer Verlag, Frankfurt, 1995.

[5] Aktion T4—Wikipedia.

The letter concluded with the following statement:

It is obvious of what tremendous importance it could be, if it were possible to bring about changes in procreative power or fertility in humans through the administration of the extract of Leopard Lily. However, this would require investigations and human experiments by an appropriately selected group of physicians ... in cooperation with the pharmacological institute of the Vienna medical faculty on inmates of the Lackenbach gypsy camp. We are fully aware that such investigations must be treated as highly sensitive state secrets... Since these considerations are initially only an idea, the fundamental correctness of which has already been established in animal experiments and the possibility of its application to humans is highly probable, the perspectives can only be hinted at, which would make the possibility of infertilizing practically unlimited numbers of people possible in the shortest possible time in the simplest possible way....[6]

As already mentioned in the previous chapter, the attempts of the second Leopard Lily research group seem to have been unsuccessful in synthesizing any active ingredients of the plant. Whether any experiments on concentration camp prisoners ever took place is uncertain.[7]

Shortly before the fall of the Third Reich, Fehringer was promoted to the rank of Obersturmbannfuehrer.[1] His fate after the war is not known.

Box 1: Office for Racial Policy (Rassenpolitisches Amt der NSDAP)[8]

- This office was founded in 1934.
- Its task was to standardize and monitor training and propaganda in all relevant areas.
- The aim was to achieve uniform regulations in the field of racial hygiene.
- By 1936, well over a thousand people had been trained as loyal speakers at the office's school of public speaking.
- The office published the newspaper "Neues Volk" (New People).
- In cooperation with Goebbel's Reich Propaganda Ministry, the office also produced several films.
- The office was most important in the early years of Nazi rule. Later, the SS and security services took over many of its core functions.
- In 1944, the office's activities were almost completely discontinued.
- After the war, it was banned by the Allied Control Council.

---

[6] Mitscherlich A, Mielke F: Medizin ohne Menschlichkeit. Fischer Verlag, Frankfurt, 1989.

[7] 'Three Million Bolsheviks Could Be Sterilised' | Project | Nuremberg. Casus pacis.

[8] Rassenpolitisches Amt der NSDAP—Wikipedia.

Box 2: Josef Bürckel (1895–1944)

- He trained as a teacher, joined the NSDAP early on, and became the regional leader of Gau Rheinpfalz in 1926.
- He was elected to the Reichstag in 1930.
- From April 1938 to March 1940, Bürckel served as the Reichskommissar for the Reunification of Austria with the Reich and was thus responsible for the integration of Austria into Nazi Germany.
- In 1938, Bürckel established the Central Agency for Jewish Emigration in Vienna which organized the forced emigration of Jews from Vienna and later played a role in their deportation and murder.
- Bürckel then served as the Gauleiter of Reichsgau Vienna from January 1939 to August 1940.
- In August 1940, he was appointed chief administrator of occupied Lorraine.
- He died on September 28, 1944; the cause of his death was said to be pneumonia, although suicide has also been suggested.

13

Oswald Pohl (1892–1951)

Oswald Pohl was a Nazi official who became the head of the Economic and Administrative Main Office. He was responsible for all administrative matters of the SS, including the vast network of concentration camps. Pohl became a key figure in the "Final Solution," as he oversaw the administration of extermination and forced labor camps. After the war, Pohl was found guilty of war crimes and crimes against humanity, sentenced to death, and executed in 1951.

Oswald Ludwig Pohl was born in Duisburg as the fifth of a total of eight siblings. After school, he joined the Imperial German Navy in 1912 and served during WW I earning the Iron Cross II Class in 1914.

He then attended courses at a trade school and began studying law and state theory at the Universität in Kiel. He dropped out of university and became paymaster for a "Freikorps," a paramilitary group that emerged after World War I, primarily composed of ex-soldiers and discontented youth.[1] In 1920, he was accepted into the Weimar Republic's new navy, the "Reichsmarine."

Pohl joined the SA in 1925 and the NSDAP in 1926. In 1933, he made the acquaintance of Himmler who was impressed with Pohl's abilities and became his mentor. Subsequently, his career was steep, and he had numerous posts and functions, too many to mention them all. He joined the SS in November 1933 and was swiftly appointed SS-Standartenführer in February

---

[1] Freikorps—Wikipedia.

© The Author(s), under exclusive license to Springer Nature
Switzerland AG 2026
E. Ernst, *The Leopard Lily Project*, https://doi.org/10.1007/978-3-032-24771-1_13

1934. His main task became the reorganization of the SS administration (Box 1).

In 1942, the "Wirtschafts- und Verwaltungshauptamt" (WVHA, Economic and Administrative Main Office) was created, and Pohl became its director. The WVHA was directly subordinate to Himmler and handled all financial and administrative matters concerning the SS as well as its vast network of forced labor industries as well as the concentration and extermination camps.[2] Pohl's role in the WVHA made him a key figure in the "Final Solution." At that stage, Pohl had become the third most influential man within the SS, after Himmler and Reinhard Heydrich.[3]

On December 12, 1942, after divorcing his first wife, Pohl married Eleonore von Brüning, widow of Ernst Rüdiger von Brüning, the son of one of the founders of the Hoechster Farbwerke, which became part of the IG Farben in 1925. The company played a significant role in the production of "Zyklon B," a pesticide originally developed for fumigation and later used by the Nazis in gas chambers to murder millions.

Pohl also played a key role in the Leopard Lily project. In his letter dated March 10, 1942, Himmler had instructed Pohl to oversee Adolf Pokorny's suggestion for sterilization experiments (Chap. 10). Subsequently, Pohl coordinated the necessary resources and gave permissions for the required procedures. Even tough Pohl's role was not hands-on, his influence and power were critical in providing the administrative framework, resources, and oversight that allowed the project to advance. His directorship of the WVHA ensured that the Nazi concentration camps were available as testing grounds whenever it should become necessary.

Whether sterilization experiments with Leopard Lily on camp prisoners ever took place is uncertain. Most sources claim that such tests were never conducted. Yet some experts disagree: "As a result of Pokorny's suggestions, experiments were conducted on concentration camp inmates to test the effectiveness of the drug."[4]

After the war, Pohl went into hiding disguised as a farm laborer. He was captured in May 1946 by the British authorities. During his interrogation, he claimed that everyone down to the lowest clerk knew what went on in the concentration camps. He was charged with war crimes, crimes against

---

[2] SS Main Economic and Administrative Office—Wikipedia.

[3] Zwangsarbeit und Vernichtung: Das Wirtschaftsimperium der SS: Oswald Pohl und das SS-Wirtschafts-Verwaltungshauptamt 1933–1945: Jan Erik Schulte, Hans Mommsen: Amazon.de: Books Schulte JE: Zwangsarbeit und Vernichtung: Das Wirtschaftsimperium der SS: Oswald Pohl und das SS-Wirtschafts-Verwaltungshauptamt 1933–1945. Schoeningh (2001).

[4] Anas GJ, Grodin MA: The Nazi doctors and the Nuremberg Code. Oxford University Press, 1992.

humanity, and membership in a criminal organization (the SS).[5] While in prison he rejoined the Catholic church—he had left in 1935—and wrote a book "Credo, Mein Weg zu Gott" (I believe. My path to God).[6] In it, he stated that "I have vigorously opposed inhumanities, insofar as I became aware of them" (Box 2).

Pohl was nevertheless found guilty of his contribution to the Holocaust, including the administration of extermination camps and the enslavement of millions. His role in the Leopard Lily project was next to immaterial to the sentence he received. Like so many of the other Nazis, Pohl claimed that he had merely been a functionary following orders and denied any personal responsibility. The tribunal rejected this defense and sentenced him to death. Pohl was executed by hanging on June 7, 1951.

Box 1: Pohl's Multiple Positions and Promotions During the Third Reich[1,7]

- June 1933: Promoted to SA-Obersturmführer.
- August 1933: Advanced to SA-Sturmbannführer.
- November 1933: Joined the SS in the rank of SS-Sturmbannführer.
- February 1934: Appointed SS-Standartenführer and chief of the administration department in the staff of the Reichsführer-SS. Began influencing the administration of concentration camps by reorganizing and professionalizing SS accounting and clerical operations.
- September 1934: Promoted to SS-Oberführer because of his success in standardizing SS accounting.
- June 1935: Appointed "Verwaltungschef" (chief of administration) and "Reichskassenverwalter" (Reich treasurer) for the SS. Initiated the "Inspektion der Konzentrationslager" (Concentration Camps Inspectorate) to oversee camp administration. Also became administrative chief over the SD Main Office and the Race and Settlement Office. Promoted to SS-Brigadeführer on the same day.
- 1935: Founded the "Gesellschaft zur Förderung und Pflege deutscher Kulturdenkmäler" (Society for the Preservation and Fostering of German Cultural Monuments), primarily dedicated to restoring Wewelsburg castle as a cultural and scientific headquarters for the SS at Himmler's request.
- Late 1930s: Appointed administrative leader of the "Lebensborn e.V.," a Nazi program promoting Aryan population growth.
- 1939: Appointed "Ministerialdirektor" in the Reich Ministry of the Interior.
- Late 1930s: Appointed chief of the "Hauptamt Verwaltung und Wirtschaft" (Main Bureau for Administration and Economy, VuWHA) within the SS.

---

[5] Pohl trial—Wikipedia.

[6] Oswald Pohl, Credo—Mein Weg zu Gott, Landshut 1950: Oswald Pohl: Free Download, Borrow, and Streaming: Internet Archive.

[7] Oswald Pohl—Wikipedia.

Box 1:

- January 1942: Himmler consolidated Pohl's offices into the "SS-Wirtschafts- und Verwaltungshauptamt" (SS Main Economic and Administrative Office, WVHA) with Pohl as its chief.
- February 1942: Assumed direct control over all concentration camps, their construction, inmate management, and economic exploitation through forced labor.
- April 1942: Promoted to SS-Obergruppenführer and General of the Waffen-SS.
- 1942: Appointed chairman of the board of directors for the "Deutscher Wirtschaftsbetrieb" (DWB, German Industrial Concern).
- March 1943: Named chairman of the board for "Ostindustrie GmbH" (Eastern Industries), an organization managing ghetto and labor camp workshops.
- July 1942–March 1943: member of the Reichstag, replacing Reinhard Heydrich.
- July 1943: "Verwaltungschef der Reichsführung SS" (Chief Administrator of the SS Leadership).
- 1944: Lost control of concentration camp administration and construction responsibilities to the Ministry of Armament.

Box 2: Oswald Pohl's Book "Credo"

- Pohl wrote this book while he was in the Landsberg prison awaiting execution.
- It is a key document in understanding the mindset of some Nazi perpetrators.
- A central theme is Pohl's conversion to Catholicism during his imprisonment, portraying it as a path to finding peace and divine forgiveness.
- The book does not provide details about the horrific crimes Pohl had committed.
- It reframed his actions as part of a general human weakness rather than specific atrocities.
- It frames his crimes in a universal context of human sin and divine grace.
- This religious rhetoric was used by several Nazis in the hope of evading personal responsibility.
- The book was part of a coordinated effort to portray war criminals in a more sympathetic light.
- Similar attempts were made, for instance, by Rudolf Höss and Albert Speer.

14

Ernst-Robert Grawitz (1899–1945)

Ernst-Robert Grawitz was a high-ranking Nazi physician who, as the "Reichsarzt" (Reich Physician) of the SS, became a central figure in Nazi medical atrocities, overseeing and approving horrific medical experiments on concentration camp inmates and playing a key part in the "Action T4." Grawitz was described as a mediocre and brutal individual who was entirely subservient to Himmler. In 1945, as the Soviet army approached Berlin, Grawitz killed himself and his family.

Ernst-Robert Grawitz was born in Charlottenburg, Berlin, into a prominent medical family. His father was the hematology professor Ernst Grawitz, and his uncle Paul Grawitz discovered a type of kidney cancer that was named after him. After serving in WWII and being released from a prisoner of war camp in 1919, Ernst-Robert studied medicine at Humboldt University in Berlin. He was involved in the Kapp Putsch[1] and became a member of the Freikorps.[2] In 1931, he joined the SS and in 1932 the NSDAP.[3] Subsequently, he made a prominent political career (Box 1).

From 1936, Grawitz served as "Reichsarzt" directly subordinate to Himmler. In this role, he became deeply involved in many Nazi atrocities

---

[1] Kapp Putsch—Wikipedia.

[2] Freikorps—Wikipedia.

[3] Ernst-Robert Grawitz | Deutsche Soldaten Wiki | Fandom.

E. Ernst, *The Leopard Lily Project*, https://doi.org/10.1007/978-3-032-24771-1_14

(Box 2), recruiting researchers to investigate chemical warfare,[4] approving, supervising, and coordinating numerous medical experiments on concentration camp inmates and facilitating the "Aktion T4" "euthanasia" program (Box 3).

In relation to the T4 program, Grawitz is reported to have said that "…it was not a pleasant task, but one must also be prepared to take on unpleasant work." He did not exclude himself from this unpleasantness and declared himself ready "… to carry out the killing of the first mentally ill person myself after the first killing centre has been set up."[5]

Grawitz was also involved in the Nazi sterilization experiments where his role was mostly that of a high-level administrator and enabler. He approved and coordinated these experiments, allocated resources, and facilitated communication between Himmler and SS doctors, like Clauberg and Schumann (Chap. 2).

After Himmler had received Pokorny's letter (Chap. 10), he wrote to Pohl (Chap. 13) asking them to get in touch with Madaus (Chap. 8) "to express his wish to not publish any further papers about medical sterilization and to offer him the option to cooperate with us – in connection with the chief medical officer SS [i.e. Grawitz] – on experiments on criminal individuals [i.e. concentration camp inmates], who need to be sterilized anyway." Himler insisted "that sterilization experiments in the concentration camps should be carried out on the basis of the plant's components."[6]

Grawitz was described as: "A mediocre, pushy, aggressive and brutal individual, entirely subservient to Himmler (…) The very type of warrant officer who sought to please his boss, who made no secret of his contempt for him, he behaved with violent authority towards his subordinates (…) the result of fanaticism and bureaucracy on a bad and untalented mind." During the Nuremberg Doctors Trial (Chap. 15), several defendants tried to shift all the blame onto Grawitz and Himmler. According to Prof Karl Gebhardt,[7] for instance, Himmler was a delusional and Grawitz narrow-minded and incompetent.[8]

When, in 1945, the Soviet Red Army closed in on Berlin, Grawitz requested permission to escape which was denied by Hitler. On April 24,

---

[4] Westemeier J, Scheib S, Uhlendahl H, Gross D, Schmidt M. Hans Wolfgang Sachs (1912–2000): From Nazi Volkstumskämpfer and chief pathologist of the Reich Physician SS to chairholder in the Federal Republic of Germany. English version. Pathologe. 2021 Nov;42(Suppl 1):11–19. doi:10.1007/s00292-019-00737-z. PMID: 33170948; PMCID: PMC8571219.

[5] Ernst-Robert Grawitz—Wikipedia.

[6] Mitscherlich A, Mielke F: Medizin ohne Menschlichkeit. Fischer Verlag, Frankfurt, 1989.

[7] Karl Gebhardt—Wikipedia.

[8] Grawitz Ernst-Robert—Mémoires de Guerre.

1945, Grawitz killed himself and his family—his wife Ilse, daughter of SS general Siegfried Taubert, and their two children—by detonating grenades under their dinner table.[5]

Box 1: Functions and Posts of Grawitz During the Third Reich

- 1935, Grawitz was appointed by Himmler as the chief medical officer of the SS and the police forces.
- 1936, deputy president of the German Red Cross (GRC).
- 1937, Grawitz aligned the GRC with Nazi ideology and prepared it for its wartime roles, in violation of traditional Red Cross principles.
- 1939, member of the leadership group responsible for the "Action T4."
- 1942, honorary professor at the medical faculty of the University of Graz.
- 1945, physician in Adolf Hitler's Führerbunker in Berlin.
- 1945, suicide.

Box 2: Some of the Worst Atrocities in Which Grawitz Was Involved

- He authorized and enabled often fatal experiments at the Dachau concentration camp, where inmates were forced into freezing water to study the effects of cold and test various rewarming methods.
- He approved the use of concentration Dachau camp prisoners in low-pressure chambers to simulate extremely high altitudes, frequently fatal experiments conducted for the Luftwaffe.
- He personally oversaw experiments at the Ravensbrück concentration camp, where female prisoners had infected wounds inflicted to test the efficacy of sulfanilamide.
- He approved operations, conducted without anesthesia, where inmates' body parts were removed at Ravensbrück to study their regeneration.
- He coordinated and approved research into methods for mass sterilization using methods like X-ray irradiation, corrosive chemical injections, and surgery (Chap. 2).
- He proposed research by Dr. Carl Værnet to find a "cure" for homosexuality, which involved hormonal and surgical experimentation on inmates at the Buchenwald concentration camp.
- He authorized experiments where subjects were shot or had limbs amputated, after taking Polygal to determine whether the drug would reduce bleeding (Chap. 5).

Box 3: The "Aktion T4"

- The Aktion T4 officially began with an order from Hitler in October 1939, which was backdated to September 1, 1939, the day WW II began.
- It was the Nazi regime's systematic program of killing people with physical and mental disabilities rooted in the Nazi ideology of racial hygiene (Chap. 1).

- Its name comes from the address of its headquarters in Berlin, Tiergartenstraße 4.
- The program initially targeted disabled children but was soon expanded to adults.
- It became a precursor to the Holocaust, using similar personal, bureaucratic methods, and gas chamber technology.
- The victims were registered, transported to special killing centers, and then murdered, primarily in gas chambers using carbon monoxide.
- The program was designed to be highly secretive, with death certificates falsified to list other causes of death.
- Yet, information began to leak out, leading to growing public protests, not least from the Catholic Church.
- In August 1941, Hitler formally ordered the program to be halted.
- However, the killings continued as "Aktion 13f14" in a decentralized fashion in hospitals and institutions for the remainder of the Third Reich.[9]

---

[9] Action 14f13—Wikipedia.

15

The Nuremberg Doctors' Trial

The Nuremberg Doctors' Trial of 1946/47 prosecuted Nazi physicians for war crimes and crimes against humanity. Its focus was on the Nazis' cruel medical experiments and on mass murder. One of the 23 defendants was Adolf Pokorny, who had proposed using Leopard Lily to sterilize millions of "Bolsheviks." The court acquitted him, because there was not sufficient evidence that his evil suggestions were ever implemented.

The Nuremberg Doctors' Trial took place between December 9, 1946, and August 20, 1947. Its purpose was to prosecute Nazi physicians and healthcare administrators for crimes against humanity. Its focus was on the various medical experiments the Nazis had performed on prisoners in concentration camps as well as the murder of patients deemed "unworthy of life" under Nazi "euthanasia" programs.[1]

Of the 23 accused, 22 were men, 20 were physicians, and three were high-ranking officials who had worked in these areas (Box 1). Karl Brandt, Hitler's personal physician, had been chosen as the lead defendant.[2] The charges against the defendants included:

---

[1] The Doctors Trial: The Medical Case of the Subsequent Nuremberg Pro... | Holocaust Encyclopedia.

[2] Nuremberg—People.

E. Ernst, *The Leopard Lily Project*, https://doi.org/10.1007/978-3-032-24771-1_15

- conducting cruel and unethical experiments on prisoners without their consent,
- involuntary sterilization of large populations,
- mass murder of patients and other individuals considered by the Nazis to be "unworthy of life."

Of the 23 defendants, 16 were eventually convicted; 7 were sentenced to death and executed in 1948, 9 received prison sentences, and 7 were acquitted.[3]

The Nuremberg Doctors' Trial was the first court case to focus on the crimes of physicians during the Third Reich. In the years that followed, more knowledge about the crimes of German and Austrian doctors would emerge. Thus, many more trials against Nazi doctors were initiated (Box 2).

The trial was also important because it established the famous Nuremberg Code. It became the basis for virtually all guidelines of medical ethics across the world. It formulated key ethical standards, most of which had previously been accepted but were agreed upon and formulated for the first time (Box 3).

One of the 23 defendants was the physician Adolf Pokorny.[4] As mentioned previously (Chap. 10), he had come across a publication by Madaus and Koch (Chap. 8) which suggested that rats could be rendered infertile by the administration of juice from the Leopard Lily. Pokorny, who had no research experience in this area, concluded that these effects would also be applicable to humans. This, he suggested, could be used to deal with a problem the Third Reich was facing. In 1941, he thus wrote to Himmler:

> If, based on this research, it was possible to produce a drug which after a relatively short time, effects an imperceptible sterilization on human beings, then we should have a new powerful weapon at our disposal. The thought alone that 3 million Bolsheviks, at present German prisoners, could be sterilized so that they could be used as laborers but be prevented from reproduction opens the most far-reaching perspectives.[5]

After the war, this letter was found in Himmler's files, Pokorny was located, arrested, and charged with conspiracy to commit war crimes and crimes against humanity as well as participation in a common plan to commit illegal

---

[3] Doctors' Trial—Wikipedia.

[4] Nuremberg—Adolf Rudolf Pokorny.

[5] Adolf Pokorny—Wikiwand.

medical experiments. War crimes specifically related to his alleged involvement in sterilization experiments. Pokony's role in proposing methods that contributed to atrocities against protected persons, i.e., forced sterilizations, prompted the charge of crimes against humanity.

On November 21, 1946, Pokorny pleaded "not guilty" to all charges. The primary evidence against Pokorny was his above-mentioned letter to Himmler, which was presented by the prosecution. The prosecution argued that the letter demonstrated intent to commit crimes against humanity. The prosecution's closing brief against Pokorny on June 16, 1947, highlighted the letter as evidence of his complicity in the Nazi sterilization program, which aimed to eliminate "enemy" populations, while exploiting their labor.[6]

Pokorny's defense argued that he had no direct involvement in conducting sterilization experiments. It further emphasized the following points:

- The Leopard Lily method had, in fact, never been tested on camp inmates.
- Pokorny's suggestions to Himmler were merely theoretical.
- Pokorny had never been a member of the NSDAP or the SS.
- He had been married to a Jew and was in no way aligned with the Nazi ideology.
- Pokorny had served humanity for 15 years as a dermatologist and was a humanitarian.
- Pokorny had always known that the proposed sterilization method could not be effective.[7]

The last point is the most intriguing one. Essentially, Pokorny's defense implied that, by writing to Himmler, Pokorny wanted to sabotage the cruel experiments aimed at mass sterilization that were, as he had been informed, being conducted in concentration camps. Therefore, the purpose of his letter was not to create but to prevent harm to camp prisoners and to populations in the occupied territories in the East.[8]

When the tribunal delivered its verdict on August 20, 1947, Pokorny was acquitted on all charges. The judges concluded that, despite Pokorny's incriminating letter to Himmler, there was insufficient evidence to prove that Pokorny had direct involvement in conducting or planning sterilization experiments. The fact that the Leopard Lily method was claimed to never have been implemented proved to be a key factor in his acquittal. The

---

6 Nuremberg—Document Viewer—Brief: prosecution closing brief against Adolf Pokorny.
7 Nuremberg—Document Viewer—Brief: Closing brief for Dr. Adolf Pokorny.
8 Mitscherlich A, Mielke F: Medizin ohne Menschlichkeit. Fischer Verlag, Frankfurt, 1989.

tribunal noted that Pokorny's proposal, while morally reprehensible, did not translate into an actionable crime within the scope of the tribunal.

The acquittal of Pokorny has remained highly controversial. Some legal experts argue that the tribunal was too lenient in cases where direct participation in crimes was not proven, despite clear ideological complicity. The acquittal also prompted debates about the tribunal's criteria for guilt: They prioritized direct involvement over intent or ideological contribution which, in the view of some experts, was unjustified.

Pokorny's letter undoubtedly aligned with Nazi genocidal goals, yet the lack of implementation and his non-membership in Nazi organizations influenced the judges' decision. The tribunal's focus on provable actions over documented intent remains a point of contention in later analyses of the Nuremberg Doctor's Trial.

After the trial, Pokorny became a free man again. He was thus allowed to continue practicing medicine. What he decided to do, where he then lived, and when he dies are questions that are unanswered to the present day.

Box 1: The Defendants at the Nuremberg Doctors' Trial

1. Karl Brandt: Physician, Reich Commissioner for Health and Sanitation. Sentenced to death by hanging.
2. Siegfried Handloser: Physician, Chief of the Medical Services of the German Armed Forces. Sentenced to life imprisonment.
3. Paul Rostock: Physician, Chief Surgeon of the surgical clinic in Berlin. Acquitted.
4. Oskar Schröder: Physician, Chief of Staff of the Inspectorate of the Luftwaffe Medical Service. Sentenced to life imprisonment.
5. Karl Genzken: Physician, Chief of the SS Medical Department. Sentenced to life imprisonment.
6. Fritz Fischer: Physician, Assistant to Karl Gebhardt at the Hohenlychen Sanatorium. Sentenced to life imprisonment.
7. Karl Gebhardt: Physician, Himmler's personal physician and Chief Surgeon of the SS. Sentenced to death by hanging.
8. Waldemar Hoven: Physician, Chief Doctor at Buchenwald Concentration Camp. Sentenced to death by hanging.
9. Joachim Mrugowsky: Physician, Chief Hygienist of the Reich and SS. Sentenced to death by hanging.
10. Helmut Poppendick: Physician, Chief of the Medical Office of the SS. Sentenced to 10 years in prison.
11. Wolfram Sievers: Administrator, Deputy Director of the Ahnenerbe Society. Sentenced to death by hanging.
12. Gerhard Rose: Physician, Vice President of the Robert Koch Institute. Sentenced to life imprisonment.
13. Siegfried Ruff: Physician, Director of the Department for Aviation Medicine at the German Experimental Institute for Aeronautics. Acquitted.

14. Hans-Wolfgang Romberg: Physician, staff member of the Department for Aviation Medicine. Acquitted.
15. Adolf Pokorny: Physician, specialist in skin and venereal diseases. Acquitted.
16. Hermann Becker-Freyseng: Physician, Chief of the Department for Aviation Medicine of the Luftwaffe. Sentenced to 20 years in prison.
17. Georg August Weltz: Physician, Chief of the Institute for Aviation Medicine in Munich. Acquitted.
18. Konrad Schäfer: Physician, Institute for Aviation Medicine. Acquitted.
19. Kurt Blome: Physician, Deputy Reich Health Leader. Acquitted.
20. Herta Oberheuser: Physician at Ravensbrück Concentration Camp. Sentenced to 20 years in prison.
21. Viktor Brack: Administrator, Chief of the Euthanasia Program. Sentenced to death by hanging.
22. Rudolf Brandt: Administrator, Himmler's Personal Administrative Officer. Sentenced to death by hanging.
23. Herman Stieve: Anatomist, head of the Anatomical Institute at the University of Berlin. Not a defendant.

Box 2: Subsequent Court Cases Against Nazi Physicians

- The Pohl Trial (1947): Tried Oswald Pohl and SS officers of the SS Economic and Administrative Office, which oversaw the concentration camp system.
- The Race and Settlement Main Office Trial (1947/48): Dealt with officials of the Race and Settlement Main Office, which oversaw the implementation of racial policies, including the "Lebensborn" program and forced sterilization.
- The Dachau Trials (1945–1948): Prosecuted staff from various concentration camps (Dachau, Mauthausen, Buchenwald, Flossenbürg, etc.).
- The Belsen Trials (1945): These trials, held by the British military, focused on staff from camps like Bergen-Belsen and Auschwitz.
- The Auschwitz Trial (1947): Tried 40 former Auschwitz staff members, including several physicians.
- The Majdanek Trials (from 1945): Tried personnel from the Majdanek concentration camp, including medical staff involved in gassings and experiments.
- The German Trials (from 1950): German courts continue to prosecute Nazi war criminals, including medical personnel. For instance, several trials in post-war Germany specifically addressed participation in the "Aktion T4" program by doctors. The Frankfurt Auschwitz Trials (1963–1965) included charges against doctors like Josef Mengele (who evaded capture).

Box 3: Key Features of the Nuremberg Code

- The human subject must give voluntary, informed consent before participating in a medical experiment.
- The experiment must be designed to yield results that are unprocurable by other means and are for the good of society.
- The experiment must be based on a thorough understanding of the problem under study to justify performing the experiment on humans.
- The experiment should be conducted such that unnecessary physical and mental suffering and injury to the subject are minimized.
- The potential benefit of the experiment to humanity must justify the risk to the individual.
- The experiment must be conducted by scientifically qualified persons who possess adequate training and skills.
- The human subject must be at liberty to bring the experiment to an end at any point.
- The scientists in charge must be prepared to terminate the experiment at any stage if they believe it is likely to result in harm to the subject.

16

Final Thoughts

The Leopard Lily project was a genocidal initiative driven by the Nazi's obsession with racial hygiene and their perceived need for large-scale, involuntary sterilizations aimed at creating a pure German people. After several series of cruel experiments on concentration camp prisoners had failed to generate a practical solution, Himmler seized on a *letter* from Adolf Pokorny which suggested that Leopard Lily could be used to secretly sterilize millions. High-ranking SS bureaucrats were then tasked with coordinating the project. Its aim was to use the plant's extracts to render the inhabitants of the occupied territories infertile. It would have allowed the Nazis to exploit their slave labor before this population would become extinct. Had Himmler's plan succeeded, the Leopard lily project would have amounted to a monstrous atrocity. Today, the project can serve as a chilling reminder of the harm that pseudoscience and the substitution of ethics with ideology can cause.

Brief Summary of the Leopard Lily Project

The Nazi ideology of race hygiene rested on an obsession with controlling human reproduction. In their effort to create a so-called pure German people, the regime considered involuntary sterilization. This prompted SS physicians to search for efficient ways to eliminate what they labeled the "Untermensch" (subhuman). However, their brutal, unethical, and often fatal experiments failed to produce a convenient method of mass sterilization which Himmler so desperately sought.

E. Ernst, *The Leopard Lily Project*, https://doi.org/10.1007/978-3-032-24771-1_16

Himmler's fascination with quackery of all types meshed seamlessly with the racial ideology he championed. Convinced that unconventional remedies might offer solutions to questions posed by the war effort, he gave physicians wide latitude to experiment on concentration camp prisoners. The camps soon became laboratories of unimaginably cruel pseudoscience, where many victims suffered and died during experiments that were as scientifically flawed as they were morally horrific.

Among the options that caught Himmler's attention was the plant *Dieffenbachia seguine*, commonly called Leopard Lily. Native mainly to South America and India, the plant's sap is loaded with needle-like oxalate crystals that can injure the skin and swell the tongue leaving a person unable to speak for days—a detail that gave the plant its unsettling nickname, "dumbcane."

The notion that Leopard Lily might be effective for sterilizing human beings had initially been suggested by Adolf Pokorny, an unknown Austrian dermatologist. In 1941 he wrote directly to Himmler, stating that the plant might be used to render "three million Bolsheviks" infertile. Himmler took some time to realize the importance of this unsolicited advice but eventually he acted sharply on Pokorny's suggestion and declared the project a scientific priority.

Pokorny's letter referred to research by Gerhard Madaus, a physician who had founded a company dedicated to plant-based medicines. Madaus approached herbal remedies with systematic rigor, studying their effects predominantly in test tubes and animal models. Among other subjects, he and his team investigated whether certain plants—including Leopard Lily—could affect fertility. Madaus and his company would later face allegations of involvement in Nazi medical crimes, but a court ultimately dismissed the accusations, even as debates around the company's wartime activities lingered.

Madaus died in 1945, and subsequently his former co-worker, Friedrich Ernst Koch, directed this research. When the results were reported to Himmler, the SS chief imagined Leopard Lily as a practical way to covertly sterilize entire populations. Koch was therefore instructed to prioritize his research on the plant. Yet, according to his post-war testimony, Koch had recognized the danger of Himmler's interest and therefore decided to boycott the investigations with a view of obstructing the Himmler's evil plans.

Even though he had alerted Himmler to Madaus' research, Pokorny did not have any further involvement in the events that followed. His letter to Himmler nevertheless placed Pokorny among the 23 defendants in the post-war Nuremberg Doctors' Trial. The judges ultimately acquitted him of crimes against humanity, finding no evidence that his cruel project of sterilizing "millions of Bolsheviks" had ever been realized.

As the Leopard Lily project proceeded, several further players became involved in it:

- Richard Rössler was a pharmacologist at the Viennese Medical School. In this role, he contributed to various unethical experiments, including aspects of the Leopard Lily project.
- Franz Fehringer, head of the Office for Racial Policy in Vienna, oversaw the persecution, dispossession, and deportation of Jews and other groups targeted by the regime. As an expert involved in the Nazis' forced-euthanasia program, he too became tied to the Leopard Lily project.
- Oswald Pohl, the influential bureaucrat who controlled the SS Economic and Administrative Main Office oversaw the entire concentration camp system, turning it into an exercise of forced labor, starvation, and murder. In addition, Himmler had made him the coordinator of the Leopard Lily project.
- Ernst-Robert Grawitz was the "Reich Physician," a man whose loyalty to Himmler far outstripped his mediocre medical ability. Grawitz approved and supervised the hideous medical experiments on camp prisoners which meant he was to have a key role in testing whether Leopard Lily would indeed prove to be useful for mass sterilizations.

In the Nuremberg Doctors' Trial of 1946–47, prosecutors confronted the world with what Nazi medicine had become: a system where ideology replaced ethics, and physicians used their skills to harm rather than heal. As the Leopard Lily project was assumed to not have caused significant human suffering, it was not high on the list of the prosecution.

Pokorny was found to be not guilty and left the Nuremberg trial as a free man. However, there is little doubt that had Himmler's plans been successful, the Leopard Lily project would have amounted to a crime as horrific as the Holocaust itself.

Timeline of the Events Leading up to the Leopard Lily Project

The events described in this book unfolded over a period of several decades. Here is an outline of those that are most relevant to the Leopard Lily project:

- June 22, 1905: Alfred Ploetz establishes the German Society for Racial Hygiene in Berlin.

- 1919: The pharmaceutical company Dr. Madaus & Co. is founded.
- 1923: Fritz Lenz is appointed to Germany's first chair in Racial Hygiene at the University of Munich.
- November 1923: The failed "Beer Hall Putsch" takes place, and Hitler is arrested and sentenced.
- April 1, 1924, to December 20, 1924: Hitler is imprisoned in Landsberg. With the help of his co-prisoner, Rudolf Hess, he writes his book "Mein Kampf."
- July 18, 1925: "Mein Kampf" is published outlining the notion of race hygiene.
- January 30, 1933: Hitler is appointed Reich Chancellor.
- April 21, 1933: Hitler appoints Rudolf Hess as his deputy.
- 1933: The Heilpraktiker Association of Germany is created.
- July 14, 1933: The Law for the Prevention of Offspring with Hereditary Diseases is passed.
- 1935: The Reich Working Group of Associations for Natural Ways of Life and Healing is established.
- September 15, 1935: The Nuremberg Race Laws are enacted.
- 1938: The Reich Working Group for Herbal Medicine and Acquisition is founded.
- 1938: The herb plantation at the Dachau concentration camp is initiated.
- 1939: Adolf Pokorny applies to join the NSDAP but is rejected due to his ex-wife's Jewish origins.
- January 23, 1939: The German Research Institute for Nutrition and Food Provision Ltd. is established.
- February 17, 1939: The "Heilpraktikergesetz" is passed regulating the new profession of lay healers.
- March 25, 1939: Dr. Gerhard Wagner, one of the most influential proponents of New German Medicine, dies; his replacement is Leonardo Conti.
- September 1, 1939: The Aktion T4 is officially started at the beginning of WWII.
- May 10, 1941: Hess flies to Scotland with the intention of negotiating peace and is declared a traitor by Hitler.
- 1941: The article "Tierexperimentelle Untersuchungen zur Frage der medikamentösen Sterilisierung" by G. Madaus and F. E. Koch is published in the journal "Zeitschrift für die gesamte experimentelle Medizin einschließlich Chirurgie."
- October 1941: Himmler receives Pokorny's letter; it claims that the plant, Leopard Lily, might be effective for inducing infertility; it takes several months until Himmler reacts.

- February 1, 1942: Oswald Pohl is appointed director of the Economic and Administrative Main Office.
- February 26, 1942: Gerhard Madaus dies.
- March 10, 1942: Himmler instructs Pohl to contact Madaus about experiments on Leopard Lily.
- June 3, 1942: Pohl reports back to Himmler about his conversation with Koch who has taken over from the late Gerhard Madaus.
- July 7, 1942: Himmler decides to prioritize the Leopard Lily project.
- October 14, 1942: A letter from Karl Gerland reaches Himmler regarding the possibility of mass sterilization using Leopard Lily in Austria.
- 1942: Sterilization experiments begin in the Auschwitz concentration camp, but Leopard Lily is not part of it.
- 1943: According to one source, unethical experiments are conducted with a Madaus preparation of Echinacea on prisoners of the Buchenwald concentration camp.
- January 12, 1945: The Auschwitz concentration camp is liberated.
- December 9, 1946: The Nuremberg Doctors Trial starts.
- August 20, 1947. Pokorny is acquitted of all charges.

Worst-Case Scenario

To illustrate the full extent of the horror that the Nazis had intended to unleash, it might be useful to conduct a thought experiment and ask: What if the Leopard Lily project had succeeded in the way Himmler had envisaged. Here is a very brief outline of this "worst case scenario":

- The Nazis win the war and occupy large parts of Eastern Europe.
- The millions living in these territories are exploited for post-war reconstructions and other tasks.
- Simultaneously, the regions are "germanized" by relocating large numbers of Germans to these areas.
- Attempts to grow Leopard Lily on a large scale are successful.
- Experiments on concentration camp prisoners confirm that Leopard Lily induces permanent infertility.
- To achieve the "germanization" within one generation, the populations of the occupied territories are forced to consume extracts of Leopard Lily in their drinking water.
- Thus, these populations are rendered infertile and become extinct.

- The genocide is completed within a matter of a few decades, and German long-term dominance is firmly established.

According to Himmler's naïve and cruel fantasy, the Leopard Lily project was aimed at the annihilation of vast numbers of humans. Even though this may seem less overtly brutal than the murderous Holocaust, its long-term effects would have been similarly devastating.

Involuntary Sterilization Outside the Third Reich

The practice of rendering infertile individuals without their free and informed consent has been employed not just in Germany. Outside the Third Reich, it was used predominantly as a public health or social welfare measure. The US might serve as an apt example.

The state of Indiana enacted the first sterilization law in 1907. It was later revised and served as a model for several other states. The US Supreme Court upheld the constitutionality of Virginia's eugenics sterilization law in a 1927 landmark case. Justice Oliver Wendell Holmes Jr. stated in this context that "Three generations of imbeciles are enough."[1]

By the 1930s, numerous US states had enacted sterilization laws, and tens of thousands of sterilizations were performed. The atrocities committed by the Nazis significantly reduced the popularity of eugenics internationally. Yet, sterilization practices continued in the US and other parts of the world based on new rationalizations.

During the 1940s, legal challenges and shifting societal attitudes began to erode the foundations of involuntary sterilization. Eventually, federal regulations aimed to ensure informed consent and to prevent abuses. Today, many US states have issued apologies, and some survivors have received compensation.[2]

The justifications used by policymakers and practitioners at the time can be summarized as follows:

- Sterilization was necessary for reducing the transmission of hereditary diseases and congenital disabilities.

---

[1] Reilly PR. Eugenics and Involuntary Sterilization: 1907–2015. Annu Rev Genomics Hum Genet. 2015;16:351–68. doi:10.1146/annurev-genom-090314-024930. PMID: 26322647.

[2] The Surgical Solution: History of Involuntary Sterilization in the United States: Amazon.co.uk: Reilly: 9780801840968: Books. Reilly P: The surgical solution: a history of involuntary sterilization in the United States. Johns Hopkins University Press (1991).

- Sterilization was a preventive health measure to lower the "burden" of genetic conditions on future generations.
- Sterilization would reduce long-term social welfare costs by preventing births among individuals considered unfit to care for children.
- Sterilization was a tool to manage population growth.
- Sterilization protected vulnerable women from sexual abuse and unintended pregnancies.

The overwhelming consensus today holds that involuntary sterilizations are unethical, harmful, and therefore unacceptable.

- They deny individuals control over their own bodies and reproductive choices.
- They undermine the principle of informed consent, a cornerstone of modern medical ethics.
- They reinforced systemic inequalities and stigmatization because marginalized groups were disproportionately targeted.
- They cause feelings of loss, grief, betrayal, and dehumanization.
- They have serious emotional and social consequences.
- They carry risks of complications and long-term health effects.
- They result in lasting mistrust in healthcare systems.

Today's human rights standards therefore reject involuntary sterilization and recognize it as a crime against humanity.[3]

The Leopard Lily Project Compared to Forced Sterilization Programs Elsewhere

It could be argued that the sterilization programs of the US (and other nations) and the Nazi plan of mass sterilization are comparable in their core violation of human rights and dignity. However, there are significant differences, e.g.:

- The Nazi's aim was to forcefully sterilize many millions of inhabitants of the vast occupied territories in the East. This amounts to orders of magnitude more victims than those sterilized in the USA.

---

[3] Maila SM, Castelyn C, Adam S. Informed consent and ethical issues pertaining to female sterilization-Scoping review. Int J Gynaecol Obstet. 2025 Jun;169(3):1037–1064. doi:10.1002/ijgo. 16100. Epub 2025 Jan 4. PMID: 39754449; PMCID: PMC12093920.

- The Nazis would have exploited the work force of their victims to prepare the land for "germanization." No such plans existed in the USA or other countries.
- The Nazi's plan was part of a genocidal program of ethnic cleansing. In other countries, forced sterilization was mainly used for public health purposes.
- The Nazi's plans were focused on racial, political, and ethnic characteristics (e.g., Jews, Roma, Slavs, "Bolsheviks"). Elsewhere, sterilization was targeted on socio-medical characteristics (e.g., "feeblemindedness," criminality, poverty).

Even if one did take the view that the Leopard Lily project was hardly more immoral than what happened in other countries, this would not render it more acceptable. No crime can be excused by the fact that elsewhere the law is being violated too!

Ethical Considerations

It could nevertheless be argued that, when the Nazis planned to sterilize millions of people living in the occupied territories, this action was—from their point of view—in the best interest of Germany. If this was true, was their sterilization program not ethical after all?

The answer is clearly negative. The Nazis' sterilization program was based on pseudoscientific eugenics, coercing individuals, violating fundamental human rights, and it served their genocidal intent (Chap. 1). While the Nazis would, of course, claim these policies to be for the "greater good" of the German state, their actions directly contradicted universally accepted core ethical principles.

As envisaged by Himmler, sterilization with Leopard Lily was a clear infraction of a person's bodily autonomy and the ethical principle of informed consent, which is a cornerstone of modern medical ethics. Involuntary sterilization permanently alters the victims' body; it is a profound violation of an individual's bodily integrity and reproductive rights.

Perhaps, the most chilling aspect of Himmler's Leopard Lily project is that it served as a proxy for genocide. The plan was:

- to sterilize millions without their knowledge,
- to keep their capacity intact for serving as slave laborers,
- to prevent future generations from being born.

This avoids murder but nevertheless amounts to genocide.

Pseudoscience and Science

It seems clear that Himmler fancied himself as a scientist. Yet, it is equally obvious that he was unable to tell science from pseudoscience. The two often seem to use similar language, and they often study similar subjects. It can therefore be easy to confuse them; in fact, this is exactly what pseudoscientists aim to achieve. But there are, of course, important differences.[4] Some of the features that render the Leopard Lily project pseudoscience include the following:

- It was based on the notion of race hygiene, itself a pseudoscientific concept.
- It was not conceived to test a hypothesis but aimed at confirming the findings of Madaus and Koch.
- It jumped to the conclusion that the animal experiments of Madaus and Koch were reproducible as well as applicable to humans.
- It lacked an understanding of the mechanism of action of the plant.
- It lacked a full understanding of the harms done by the plant.
- Its aim was not to benefit but to harm humans and thus it was deeply unethical.

Science and pseudoscience diverge particularly sharply in their ethical and moral foundations. While science is built upon principles of honesty, openness, and responsibility, pseudoscience undermines these values often by placing ideology and belief over evidence and truth. Science is not least an ethical enterprise, and the divide between science and pseudoscience is a matter of profound moral importance. The ethical stakes become especially acute when pseudoscience causes harm or like in the Leopard Lily project is meant to cause harm. Even when pseudoscientists act with conviction, the lack of accountability magnifies the potential for damage. In essence, pseudoscience is an abuse of science.

At the heart of the Leopard Lily project was dangerous pseudoscience combined with fanatical ideology, immorality, and political power. Such an unholy alliance would inevitably lead to disaster. That the planned genocide did not happen was simply owed to the circumstances. Similar disastrous

---

4 Science versus pseudoscience

alliances are not exclusively phenomena of the Third Reich but have since occurred with tragic regularity. Three recent examples must suffice:

- Trofim Lysenko's rejection of Mendelian genetics was enforced by Joseph Stalin and the Soviet Communist Party, leading to the persecution and execution of numerous genuine geneticists and contributing to massive crop failures and subsequent famines resulting in millions of deaths.[5]
- Under Thabo Mbeki, the South African government adopted an AIDS denialist stance and restricted the rollout of life-saving anti-retroviral drugs as well as programs for mother-to-child HIV transmission prevention. These policies resulted in the deaths of over 330,000 South Africans and thousands of preventable HIV infections in babies.[6]
- The current anti-vaccine ideology of US officials is leading to a sharp decline in childhood vaccinations. Public health experts therefore warn that this trend will contribute to more disease and significantly increased vulnerability to future pandemics.[7]

The story of the Leopard Lily project can serve as a reminder of the dangers caused by unholy alliances of pseudoscience, ideology, immorality, and political power. These dangers have not ended with the Third Reich. If the book can contribute to reducing the risks of future recurrences, it was worth the effort of writing it.

---

[5] Trofim Lysenko—Wikipedia.

[6] HIV/AIDS denialism in South Africa—Wikipedia.

[7] Robert F. Kennedy Jr.—Wikipedia.

Index

GPSR Compliance
The European Union's (EU) General Product Safety Regulation (GPSR) is a set
of rules that requires consumer products to be safe and our obligations to
ensure this.

If you have any concerns about our products, you can contact us on

ProductSafety@springernature.com

In case Publisher is established outside the EU, the EU authorized
representative is:

Springer Nature Customer Service Center GmbH
Europaplatz 3
69115 Heidelberg, Germany